Heat Transfer from Flames

W. A. Gray,
B.Sc., Ph.D., C.Eng., F.Inst.Gas E., F.Inst.F.
Department of Fuel and Combustion Science, University of Leeds

J. K. Kilham,
B.Sc., Ph.D., C.Chem, F.R.I.C.
Department of Fuel and Combustion Science, University of Leeds

R. Müller,
B.Sc., Ph.D., M.Inst. F.
Formerly at the Department of Fuel and Combustion Science,
University of Leeds

UWIST

AF740438

Elek Science London
and New Hampshire USA

QC 320.2 .G7

X

© Paul Elek (Scientific Books) Ltd., 1976
First published in Great Britain in 1976 by
Paul Elek (Scientific Books) Ltd.,
54–58 Caledonian Road,
London N1 9RN

and in USA by
Paul Elek Inc.,
10 South Broadway,
Salem, New Hampshire 03079

ISBN 0 236 40044 4

UWIST
CARDIFF
library
10 NOV 1976
ACCESSION NUMBER 104453
CLASS NUMBER 621.4022
GRAY

Printed in Great Britain by Unwin Brothers Limited
The Gresham Press, Old Woking, Surrey, England.
A member of the Staples Printing Group.

Heat Transfer from Flames

20 0104453 6 TELEPEN

L1

Contents

Acknowledgement

We are very grateful for the help given to us during the preparation of this text by Mr. Alec Taylor.

W.A.G.
J.K.K.
R.M.

1 Introduction

Calculations of the heat transfer from flames and prediction of the thermal performance of combustion systems are of great importance, particularly with regard to fuel economy. The results of these calculations can be used either to improve the thermal efficiency of existing plant or to design new plant.

Flames and combustion products are defined in this introductory chapter together with a discussion of those components of flames and products which are involved in the processes of heat transfer. Subsequent chapters are concerned with basic convective and radiative heat transfer, convective and radiative heat transfer from flames and, finally, heat transfer from flames in practice.

It is difficult to define a flame precisely but, with some exceptions, a flame can be said to possess four main attributes. Flames are associated with (1) oxidation processes, (2) highly exothermic reactions between gases, (3) rapid temperature rise, and (4) emission of light. The reactants and products of the oxidation processes are present both in the flame and beyond the flame itself as defined by the outer envelope which emits light. Oxidation may, however, be complete at this envelope and virtually no reactants appear beyond it. In any case, heat transfer is important, not only from the flame, but also from the products of the oxidation or combustion reactions. In the text which follows, the term 'flame' will be used to include the products which arise from it, unless it is obvious that only processes taking place within the flame envelope are being considered.

The most familiar, and the most industrially significant, flames are those arising from the combustion of a hydrocarbon fuel whether it be solid like coal, liquid like many of the petroleum fuels or gaseous like natural gas. As far as heat transfer is concerned, it is appropriate to consider the gaseous and particulate components of the products of combustion separately.

1.1 GASES

A simple hydrocarbon fuel burns to produce carbon dioxide and water, if combustion is complete.
In oxygen:

$$C_nH_m + (n + m/4)O_2 \rightarrow n\ CO_2 + m/2\ H_2O$$

In air:

$$C_nH_m+(n+m/4)(O_2+3.76N_2) \rightarrow nCO_2+m/2H_2O+3.76(n+m/4)N_2$$

The concentration by volume of carbon dioxide and water vapour in the products of complete combustion for three fuels are given in Table 1.1.

Table 1.1 shows that the concentration of water vapour in the case of methane is twice that of the carbon dioxide, whereas, for the petroleum fuel oils, the concentration of water vapour is approximately equal to that of the carbon dioxide. These simple ratios are convenient to use in some calculations of radiative heat transfer from flames (Section 5.1). The carbon-hydrogen ratios of coals vary considerably and it is not possible to adopt this technique for coals in general.

If the temperature of the products is high enough, various dissociated species will also be present in significant amounts. For example, the combustion products of the fuel in air may contain CO, H_2, O_2, H, O, OH, N, NO_2 and NO in significant concentrations if the temperature is high enough, whilst the flame

Table 1.1. Volumetric concentration of carbon dioxide and water vapour in the combustion products of three fuels.

Fuel	Excess air %	CO_2 %	H_2O %
Coal (carbon, 74.0 % wt; hydrogen, 5.5 % wt;	0	16.8	8.2
oxygen, 8.6 % wt; sulphur, 1.2 % wt;	20	14.1	6.9
nitrogen, 1.0 % wt and water, 5.0 % wt)	30	13.1	6.4
Medium fuel oil (carbon,	0	14.1	11.1
85.3 % wt and hydrogen,	20	11.8	9.3
11.2 % wt)	30	11.0	8.6
Methane	0	9.5	19.0
	20	8.1	16.1
	30	7.5	15.0

may contain in addition, various species such as CH or CH_2 which arise during combustion. The equilibrium composition of the products may be calculated by the kind of technique described in Appendix 1.

1.2 PARTICLES IN COMBUSTION PRODUCTS

In addition to the gaseous species, both liquid and solid particulate material may be present in flames, although in those flames derived from gaseous fuels, the only solid material which can be present is finely divided carbon. In flames derived from liquid and solid fuels, larger particles are formed from the droplets or particles of fuel. These may persist in the combustion products.

1.2.1 Finely Divided Carbon or Soot

The finely divided carbon particles formed in the flame cause the flame to be luminous. If combustion of these particles is not completed within the flame, they will emerge from the flame as soot. Soot emanating from, and the carbon within, the flame may be sampled and studied under the electron microscope. The particles are roughly spherical, of 0.005 μm to 0.25 μm (50 to 2500 Å) in diameter, and clustered together in chains as shown in Figure 1.1. Diffusion flames and slow-burning premixed flames produce larger particles, whilst smaller particles are produced by fast-burning premixed flames.

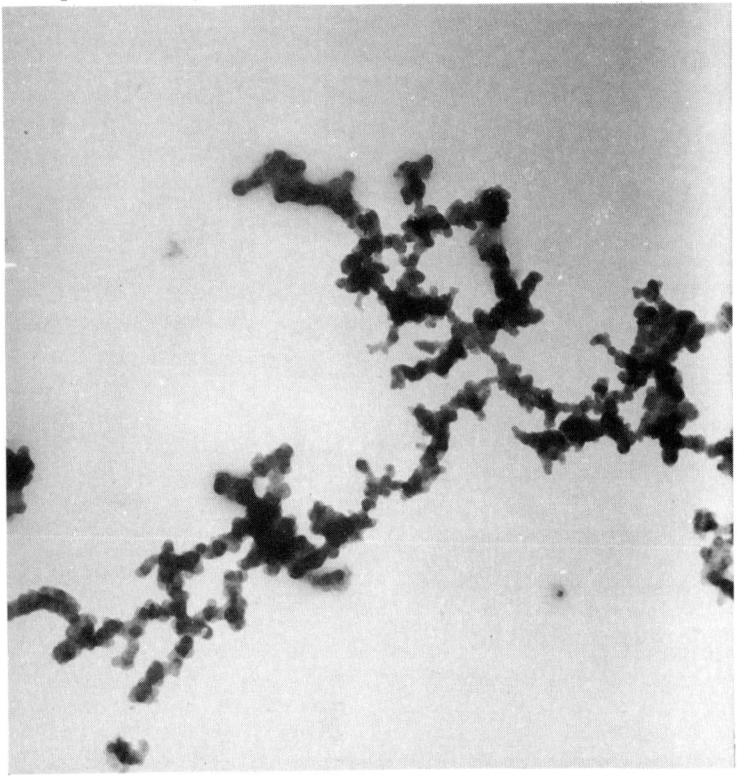

Figure 1.1 Electromicrograph of soot

The structure of the soot particles is important in that it influences their optical properties which control the radiative transfer to and from the soot. The particles in most soots have a basically graphitic structure with each particle consisting of a number ($\simeq 10^3$) of small crystallites of around 2 nm in size, although there is a variation in soots from different sources. Thomas (1962) quotes sizes up to 4.8 nm for acetylene black and compares these with crystallites of a polybenzenoid hydrocarbon such as circumanthracene (C_{40} H_{16}). The ratio of hydrogen to carbon atoms in circumanthracene is 1:2.5 but the ratio for some other soots has been shown by chemical analysis to be 1:8.

The production of soot depends on two factors, formation of carbon particles and their combustion in the flame, and they may very well be different in different types of flames, including the broad classification of diffusion and premixed flames.

Diffusion flames

A simple diffusion flame usually has a pale bluish base and is luminous over the remainder of its length. Sometimes, soot is formed at the tip of the flame. Some combustible gases do not form carbon in their flames; they include carbon monoxide, methyl alcohol, formaldehyde, formic acid, carbon disulphide and cyanogen.

Very small diffusion flames are not luminous and carbon is only produced when the height of the flame is increased above a critical value. This critical height varies with the diameter of the burner or type of wick (for vapours) and the type of fuel, even for different members of an homologous series. The tendency to produce carbon increases with molecular weight for paraffins and alcohols but decreases for olefins, di-olefins, aromatics and cyclo-paraffins. The carbon to hydrogen ratio is obviously significant, but so too is the structure of the molecule; for example, the more highly branched paraffins have a greater tendency to produce carbon.

The luminosity of the flame depends on the amount of carbon formed and its temperature. If flame temperatures are the same, as they are for many hydrocarbons, the onset of luminosity is a convenient index of carbon formed. In fact, the index used in the early experiments was the height at which soot appeared at the tip of the flame, or the tendency to smoke, as it was called. The onset of luminosity is, of course, a better index because the tendency to smoke depends on the passage of carbon through the flame.

The luminosity of the carbon in the flame and its liberation as soot or consumption by oxygen, depends on the position at which it is formed. Two examples illustrate this point. Methane requires a high temperature for decomposition, so carbon is formed in a region of relatively high temperature and late in the flame where oxygen is present. The particles will be hot and luminous but should also be quickly consumed so that little soot is produced. On the other hand, acetylene decomposes at lower temperatures and carbon is formed in a relatively cool region of the flame in the presence of lower quantities of oxygen. Soot is more likely to be produced. The tendency for carbon and soot formation increases with gas flow but, if the flame becomes turbulent, this tendency is offset by entrainment of more air into the flame.

Premixed flames

If the ratio of oxygen atoms to carbon atoms in a premixed flame is greater than 1 ($O/C > 1$) it seems reasonable to assume that all the carbon goes to form carbon monoxide and no solid carbon will be present. If $O/C < 1$, solid carbon must be expected.

In practice, premixed acetylene-air flames become luminous when O/C is reduced to 1.2, but for many fuels the ratio is even higher - 1.5 to 2.5 - as shown by the data of Street and Thomas (1955) in Table 1.2.

Effects of pressure and temperature on soot formation

By and large, soot formation is greater at higher pressures but, in many cases, the effect is small for premixed flames.

Higher flame temperatures tend to shift the

Table 1.2 Onset of carbon in premixed flames

(Street and Thomas, 1955)

Fuel	Oxygen/carbon ratio at the point of carbon formation
Ethane	2.1
Propane	2.1
n-pentane	2.2
n-octane	2.2
Ethylene	1.67
Acetylene	1.2
Benzene	1.75
Ethyl alcohol	2.5
Acetaldehyde	1.9
Diethyl ether	2.0

critical ratio to slightly richer mixtures because a greater effect is produced on the oxidation reactions rather than on the reactions producing carbon. The reverse is true for very rich mixtures and greater quantities of soot are produced at high temperatures.

Effects of additives on soot formation

The majority of substances, when added to premixed flames in small quantity, have little effect on soot formation (Gaydon and Wolfhardt, 1970). One exception is sulphur trioxide which, even in small quantities, increases the luminosity of premixed towns' gas/air and isobutane/air flames, in spite of the fact that the oxygen/carbon ratio is increased by the addition. Sulphur trioxide has little effect on diffusion flames.

The addition of 45% CO_2 stops carbon formation in a methane/air diffusion flame (Arthur, 1950), in accordance with the Boudouard reaction, which is

$$CO \rightleftharpoons \tfrac{1}{2}\,CO_2 + \tfrac{1}{2}\,C_{solid} + 20\ \text{kcal mol}^{-1}$$

Carbon formation is also reduced by some metal salts and barium salts are used to reduce smoke from diesel engines.

Electrical effects

Soot formation may be changed significantly by the application of electrical fields. Weinberg (1968) has shown considerable reductions in the amount of carbon formed and in the size of particle produced when a strong field, in the region of 10 kV, is applied. In this case, the structure of the soot was amorphous.

Production of soot

It is not possible to predict the amount of solid carbon present in a flame from equilibrium compositions. The rates of the reactions leading to the formation and removal of carbon determine whether carbon will be present. The formation of soot can be stated simply to begin with a small organic molecule and to end with a relatively large solid particle containing many thousands of atoms and a much higher carbon/hydrogen ratio than the original molecule. Such a transition must involve dehydrogenation and growth or condensation. A number of possible routes involving these processes have been suggested (Table 1.3) but no single one has been accepted in preference to the others.

1.2.2 Large Particles

(a) *Char or carbonaceous residues*

Coal particles decompose, when heated to a sufficiently high temperature, to produce volatile matter which is a mixture of combustible gases, carbon dioxode and water vapour. The residue, after evolution of the volatile matter, is a char particle which is comparable in size to the original coal particle. The physical form of this char is often

Table 1.3 Suggested routes for the formation of soot.

Route	Comments
Dehydrogenation and polymerisation via acetylene	Acetylene is a stable intermediate in rich premixed flames of other hydrocarbons. It gives soot readily under appropriate conditions, presumably because of its unsaturated character; but on the basis of O/C ratio, acetylene has a very low tendency to form soot.
Polymerisation via aromatics and polycyclic hydrocarbons	The evidence for this route lies in the similarity of the benzene ring to that of graphite and the ease with which benzene and naphthalene produce soot, both in premixed and diffusion flames. The drawback to this hypothesis is the lack of direct evidence of the presence of aromatics as intermediaries found in the flame.
Condensation and graphitisation	The formation of a mist has been observed in the pyrolysis of methane and other hydrocarbons (Parker and Wolfhard, 1950). If the products of the pyrolysis are burned as a diffusion flame, it is possible for the mist droplets (polymers) to graphitise as they pass through the flame. This route, however, is not compatible with the effect of electric fields on diffusion flames, which indicates that some growth of carbon particles must occur.
Nucleation, particle growth and coagulation	The growth of carbon particles held in a furnace through which hydrocarbon vapour is passed has been demonstrated by Tesner (1959) and it has been suggested that similar growth on appropriate nuclei, such as ions or C_2, may take place in flames. The clustering of the spherical particles in soot, as shown by electron micrographs, seems to be evidence for a coagulation process.

that of a hollow sphere called a cenosphere. In these cases, where the mass is concentrated in the shell, it is suggested that the char burns effectively at constant 'diameter' until a very late stage in the combustion process.

Coals fed to power station furnaces in the UK are ground so that approximately 80% is less than 75 µm (200 B.S. mesh). The size distribution is often expressed in terms of the mass fraction of material (dm) within an infinitesimally small interval of size (dd) by the Rosin-Rammler law:

$$\frac{dm}{dd} = n\ b\ d^{n-1}\ \exp(-b.d^n) \qquad (1.1)$$

where d is the particle size and b and n are constants. Integration of this expression gives the weight percentage (R) of material of particle size greater than d:

$$R = 100\ \exp(-b.d^n) \qquad (1.2)$$

It is convenient to express this function in the form:

$$R = 100\ \exp\left[-(d/d')^n\right] \qquad (1.3)$$

because d' is approximately equal to the mean particle size on a weight basis for the values of n found with ground coals (Field, Gill, Morgan and Hawksley, 1967). Values of $n = 1.2$ and $d' = 50$ µm may be used as reasonable estimates for pulverised coal (pulverised fuel or p.f.) in the UK.

Combustion of heavy residual fuel oils produces coked particles which result from the heavy bituminous material present. Again, it is usually assumed that these carbonaceous residues are approximately the same size as the original droplets in the spray and that the major part of the time taken for the combustion of the fuel is due to the combustion of these residues.

The droplet size distribution of a spray may be represented by the Nukiyama-Tanasawa equation (Equation 1.4), which is similar to that of the Rosin-Rammler expression (Equation 1.1). The Nukiyama-Tanasawa expression is usually stated in terms of the number (dN) of particles within an infinitesimally small interval of size (dd), as follows:

$$\frac{dN}{dd} = a\, d^{\alpha} \exp(-b\, d^{\beta}) \qquad (1.4)$$

where a, α and β are constants. The usual values for sprays produced from particular atomisers are given in Table 1.4 and describe sprays with droplet sizes in the range 0 to 200 μm.

Table 1.4 Constants for the Nukiyama-Tanasawa Equation

Type of atomiser	value of α	value of β
Swirl	0.5	1
Impinging jet	1	1
Pressure jet	2	1
Twin-fluid	1	1

The use of a number distribution for liquid sprays, rather than a weight distribution, stems from the practical consideration that the size distribution of sprays is usually determined experimentally on a number basis and that of solid particles on a weight basis.

(b) Ash

On combustion, the mineral matter in heavy fuel oil and coal becomes ash. In fact, the quantity of mineral matter in a fuel is reported in terms of the ash produced after combustion under standard conditions in the laboratory. Petroleum fuel oils are characterised by ash contents of about 0.1%, whereas coals have a much greater amount of ash associated with them. For example, pulverised coals in the UK have ash contents in the range 15 to 25% by weight (Littlejohn, 1966). Some of the mineral matter is separated during grinding to produce 40-70% clean coal particles which contain only inherent mineral matter (equivalent to less than 2% of ash and 20-40% of particles comprising coal and mineral matter; the remainder comprises discrete mineral particles. After combustion, the ash particles take a number of forms depending on the temperature history to which they have been subjected. The sizes of these particles cover the same range as the coal particles, but Figure 1.2 shows that more ash is derived from the smaller coal particles than the larger.

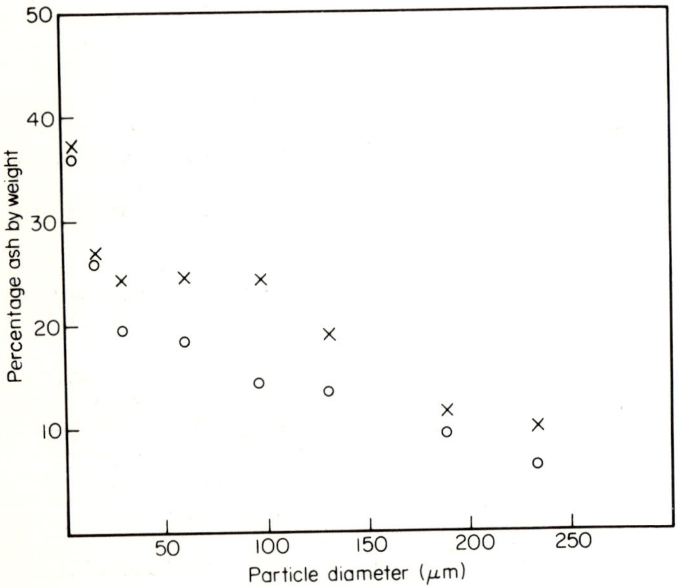

Figure 1.2. Ash in pulverised fuel used in the UK (after Littlejohn, 1966)

2 Convective Heat Transfer

Convective heat transfer is the process by which heat is transferred from one point in a fluid to another point in the fluid or to a wall by movement of the fluid. The process is one of both conduction and movement of the fluid, but the single term convective heat transfer is usually used to encompass both.

Fluid movement can be induced by two mechanisms. Free or natural convection occurs as a result of density differences arising from temperature variations in the fluid. When the motion is caused by mechanical means such as a blower, the process is known as forced convection.

Experience has shown us that the main resistance to heat flow from a surface to a fluid, or vice versa, is in the comparatively thin region adjacent to the wall surface. Almost all the temperature gradient between the fluid and the wall occurs within this region which is called the thermal boundary layer. The calculation of rates of convective heat transfer to or from a surface often makes use of the analogous concept of a velocity boundary layer in which the velocity varies from zero to the main stream velocity. Once the temperature gradient ($\mathrm{d}T/\mathrm{d}y$) in the fluid adjacent to the wall is determined, heat flow at the wall is assumed to take place by conduction through the boundary layer and not by convection, that is,

$$Q = k\,A\,(\mathrm{d}T/\mathrm{d}y) \qquad (2.1)$$

where k is the thermal conductivity of the fluid and A is the area of the surface. The thickness and nature of this boundary layer is determined by the motion of the fluid. Since convective heat transfer is so closely linked with fluid motion or flow, it is necessary to know something about the mechanism of flow before rates of heat transfer can be determined. In particular, it is important to establish whether the flow is laminar or turbulent and, as a consequence, the nature of the boundary layer.

In laminar, or streamline flow, the fluid moves in layers, each following a smooth and continuous path. In turbulent flow, the path of any small volume element is irregular although statistically the overall motion of the fluid is regular and predictable. If the flow is laminar along a surface at a different temperature, heat is transferred only by molecular conduction within the fluid as well as at the interface between the fluid and the surface. In turbulent flow, however, the conduction mechanism is modified and augmented by eddies which carry fluid across the streamlines. This fluid acts as a carrier of heat which is transferred by mixing with other fluid particles. An increase in the rate of mixing (or turbulence) will also increase the rate of convective heat flow.

2.1 THE BOUNDARY LAYER

Flow of Fluid Past a Flat Plate

When a fluid flows along a surface, the fluid in the vicinity of the surface is slowed down by viscous forces so that the velocity adjacent to the surface is zero. The thickness of the fluid layer, in which the velocity varies from zero to the main stream velocity u_s, is called the hydrodynamic boundary layer. If the flow of fluid approaching the leading edge of the plate is laminar, the boundary layer usually grows in thickness as the distance along the surface increases from the leading edge, as shown in Figure 2.1.

The shear stress (s) at any point in the $x - z$ plane is given as:

$$s = \mu(du/dy) \qquad (2.2)$$

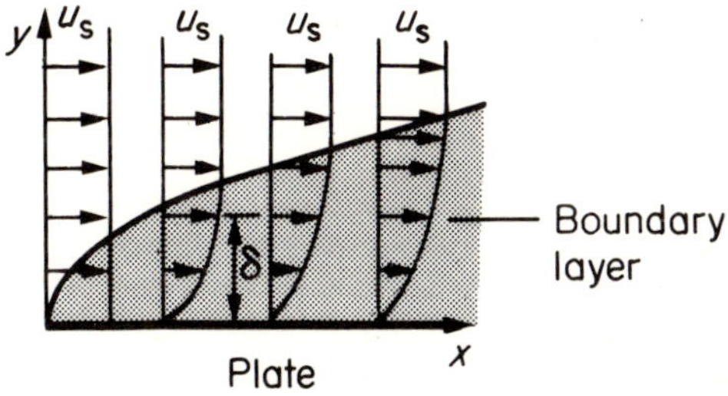

Figure 2.1 Development of laminar boundary layer on a flat plate

The velocity profile in the laminar boundary layer can be described approximately by the cubic equation:

$$u/u_s = 3y/\delta - 0.5(y/\delta)^3 \tag{2.3}$$

On the basis of this expression, the thickness of the boundary layer may be shown to be:

$$\delta = 4.64\ x/Re^{0.5} \tag{2.4}$$

where *Re* is the Reynolds number based on x, the distance from the leading edge of the plate, that is, $\rho\ u_s\ x/\mu$. In addition, it is useful to determine the losses due to friction, f, which arise from flow over the plate. This is achieved as follows:

The shear stress at the wall (s_w) is expressed in terms of the dynamic head of the free stream:

$$s_w \propto \rho\ u^2/2$$
$$= f \rho\ u^2/2 \tag{2.5}$$

where f is a dimensionless friction factor.

Equations 2.3, 2.4 and 2.5 lead to a value of f:

$$f = 0.647/Re^{0.5} \qquad (2.6)$$

At a critical distance (x_c) from the leading edge layer, the flow pattern changes from laminar to turbulent. Eddies and vortices form and destroy the laminar regularity of the boundary layer motion and only a very thin layer of fluid in which laminar motion persists remains near the surface. This portion of a turbulent boundary layer is called the laminar sublayer. Between the fully turbulent region and the laminar sublayer an intermediate region exists which is called the buffer layer. In this region the flow cannot be classed as either turbulent or laminar. The structure of the flow in a turbulent boundary layer is shown schematically on an enlarged scale in Figure 2.2.

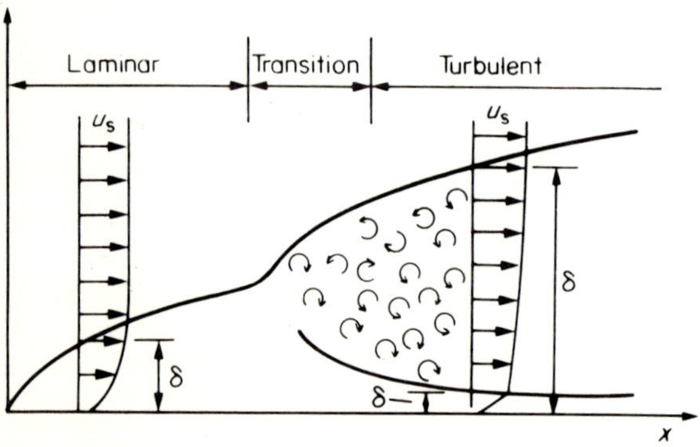

Figure 2.2 Formation of the turbulent boundary layer

The critical distance (x_c) has been determined experimentally to be:

$$x_c = 5 \times 10^5 \, \mu/\rho \, u_s \qquad (2.7)$$

In other words, the critical distance occurs when the Reynolds number is equal to 5×10^5. If the free stream approaching the plate is turbulent, a laminar boundary layer still occurs, although the critical distance is less than in the laminar case.

The friction factor for a turbulent boundary layer cannot be obtained theoretically; experimental data have, however, provided the following correlation:

$$f = 0.0592/Re^{0.2} \qquad (2.8)$$

Based on the Prandtl statement that the velocity profile in the turbulent boundary layer obeys the relationship,

$$u/u_s = (y/\delta)^{1/7} \qquad (2.9)$$

the thickness of the turbulent boundary layer is given as:

$$\delta = 0.38\ x/Re^{0.2} \qquad (2.10)$$

Flow of Fluid through a Cylindrical Tube

When a fluid approaches a tube with uniform velocity, a boundary layer gradually builds up until it reaches the centre of the tube. At this stage, the velocity profile is fully developed and does not change along the tube (Figure 2.3). In practical terms, laminar flow, which has a parabolic velocity profile (Equation 2.12), is fully developed at a distance ℓ from the entrance to the tube where,

$$\ell/D = 0.0288\rho\ u_m D/\mu = 0.0288\ Re \qquad (2.11)$$

where D is the diameter of the tube, u_m the mean velocity in the tube and μ the viscosity of the fluid.

The velocity in laminar flow at a distance y from the wall of a tube of radius r is:

$$u/u_r = 2y/r - (y/r)^2 \qquad (2.12)$$

where u_r is the velocity at the axis of the tube and the mean velocity is:

$$u_m = u_r/2 \qquad (2.13)$$

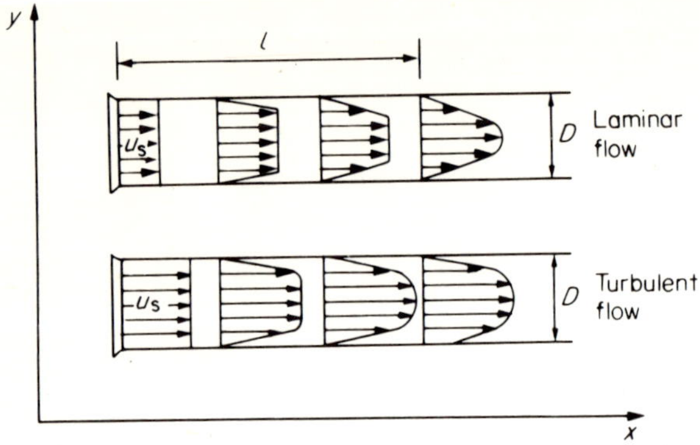

Figure 2.3 Fully developed velocity profile in a tube

Using Equations 2.2, 2.5 and 2.12, the friction factor (f) for laminar flow is given as:

$$f = 16/Re \tag{2.14}$$

a law which has been confirmed experimentally.

The friction factor for fully developed turbulent flow cannot be deduced theoretically; it has been derived experimentally as:

$$f = 0.08/Re^{0.25} \tag{2.15}$$

Equations 2.14 and 2.15 are usually presented in graphical form as shown in Figure 2.4.

2.2 CONVECTIVE HEAT TRANSFER

The rate of heat transfer by convection between a fluid and a surface may be represented by:

$$q = h_c A \Delta T \tag{2.16}$$

where q is the rate of heat transfer, A is the area through which the heat is transferred and ΔT is a temperature difference (usually the difference between the surface temperature and the temperature of the fluid some distance from the surface).

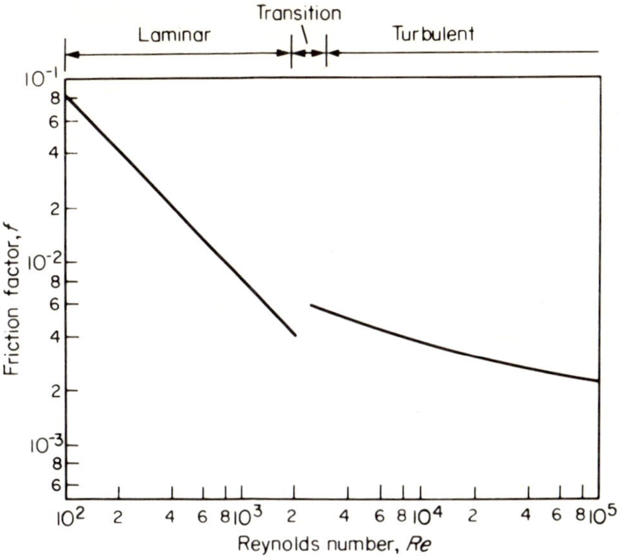

Figure 2.4 Friction factor chart for a smooth tube

Equation 2.16 is best regarded as a definition of h_c, which is a constant of proportionality, chosen so that, once the area is known and the temperature difference selected, Equation 2.16 becomes valid. The proportionality constant is usually called a heat transfer coefficient. In fact, h_c is not a constant but depends on the geometry of the surface, the velocity and physical properties of the fluid and frequently upon the value of the temperature difference ΔT.

Thus, the calculation of convective heat transfer from a fluid to a surface involves the derivation or measurement of the coefficient h_c based on an appreciation of the motion of the fluid in the vicinity of the wall.

2.3 EVALUATION OF CONVECTIVE HEAT-TRANSFER COEFFICIENTS

The calculation of convective heat transfer from a fluid to a wall may be accomplished in one of three ways: (1) by analysis of conductive heat transfer through the boundary layer either by exact mathematical solution or by approximate integral methods; (2) by using the analogy between heat and momentum transfer, the rates of heat transfer may be predicted from measurements of the shear stresses between a fluid and a wall and (3) by the direct measurement of heat transfer in a limited number of experimental situations and their wider application based on dynamic similarity and the use of dimensional analysis.

Analytical solutions to problems of forced convective heat transfer can be found for laminar flow, although they are difficult even for simple geometric shapes. The Reynolds analogy provides a semi-empirical approach to turbulent convection, but is of limited application. The direct measurement of heat transfer coefficients, coupled with correlation on the basis of dimensionless groups is still the most important method.

2.4 DIMENSIONAL ANALYSIS

Dimensional analysis is a technique for expressing the behaviour of a physical system in terms of the minimum number of dimensionless groups of the physical variables which control that behaviour. The first step is to decide what those physical variables are.

(a) Forced convection

Theoretical considerations and experimental results indicate that the heat transfer coefficient for forced convection is a function of the following variables:

μ , the viscosity of the gases

k , the thermal conductivity

c_p, the specific heat

ρ , the density

U , the mean gas velocity

L , a characteristic linear dimension of of the solid in contact with the fluid

That is,

$$h_c = f(\mu, k, c_p, \rho, U, L) \qquad (2.17)$$

The basic mathematical principle of dimensional analysis is that, in any physical equation, both sides of which can be written as power functions or as the algebraic sums of power functions, the sum of the exponents of the basic units must be the same on the left and right sides. Thus, if Equation 2.17 can be written in the form,

$$h_c = A\ \mu^a\ k^b\ c_p^{\ c}\ \rho^d\ U^e\ L^f \qquad (2.18)$$

where A, a, b, c, d, e and f are constants, the dimensions on the right hand side of Equation 2.18 must be the same as h. These dimensions are expressed in terms of selected primary dimensions which, for the present purpose can be mass (M), length (L), temperature (T) and time (t).

As an example, the variables involved in correlating experimental convective heat transfer data for gases flowing over a tube, together with their primary dimensions are given in Table 2.1.

Table 2.1. Variables involved in convective heat transfer from gas to a tube

Variable	Symbol	Dimensional equation
Tube diameter	D	L
Thermal conductivity of gases	k	ML/t^3T
Mean gas velocity	U	L/t
Density of gases	ρ	M/L^3
Dynamic viscosity of gases	μ	M/Lt
Specific heat at constant pressure	c	L^2/t^2T
Heat-transfer coefficient	h_c	M/t^3T

Equating the dimensions in Equation 2.18:

$$M/t^3T = (M/Lt)^a (ML/t^3T)^b (L^2/t^2T)^c (M/L^3)^d (L/t)^e (L)^f$$

mass:	1	$=$	a	$+ b$		$+ d$		
length:	0	$=$	$- a$	$+ b$	$+ 2c$	$- 3d$	$+ e$	$+ f$
time:	$- 3$	$=$	$- a$	$- 3b$	$- 2c$		$- e$	
temperature:	$- 1$	$=$		$- b$	$- c$			

There are six unknowns in these equations and a solution is only possible if two of them are given values. Choosing those which correspond to the specific heat and the velocity gives:

$$a = c - e$$
$$b = 1 - c$$
$$d = e$$
$$f = e - 1$$

Equation 2.18 can now be written as:

$$h_c = A \frac{k}{D} \left(\frac{\rho\ U\ D}{\mu}\right)^e \left(\frac{c_p\ \mu}{k}\right)^c \qquad (2.19)$$

or,

$$\frac{h_c D}{k} = A \left(\frac{\rho\ U\ D}{\mu}\right)^e \left(\frac{c_p\ \mu}{k}\right)^c \qquad (2.20)$$

The dimensionless group, $h_c\ D/k$ is known as the Nusselt number (*Nu*), $\rho\ U\ D/\mu$ is a Reynolds number (*Re*), with the tube diameter as the characteristic length, and $c_p\ \mu/k$ is called the Prandtl number.

Nusselt number

$$Nu = h_c\ L/k \qquad (2.21)$$

The Nusselt number can be interpreted physically as the ratio of the temperature gradient in the fluid immediately in contact with the surface to a reference temperature gradient.

$$Nu = \frac{(dT/dy)_{y=0}}{(T_f - T_s)/L} \qquad (2.22)$$

where L is a characteristic length of the object to which the heat flows, y is the distance normal to the surface, and T_s and T_f are the surface temperature and the temperature at a considerable distance from the surface, respectively.

In practice, the Nusselt number is a convenient measure of the rate of convective heat transfer because either h_c can be calculated or the rate of heat transfer determined from a variant of Equation 2.16, that is,

$$q = Nu\ A\ k\ \Delta T/L \qquad (2.23)$$

Reynolds number

$$Re = U\ \rho\ L/\mu \qquad (2.24)$$

The Reynolds number can be shown to be the ratio of inertia forces to viscous forces in a fluid and its significance is, that in geometrically similar systems, the velocity distributions will be similar for a given value of Reynolds number. In long tubes, for example, the flow is laminar when $Re < 2300$ and fully turbulent for $Re > 10\ 000$. Transition from laminar to turbulent flow takes place in the range of Reynolds numbers between 2300 and 10 000.

Prandtl number

$$Pr = c_p\ \mu/k \qquad (2.25)$$

The prandtl number is a function of the fluid properties alone and can also be expressed as the ratio of the kinematic viscosity (ν) to the thermal diffusivity as follows:

$$Pr = \frac{\mu/\rho}{k/\rho\ c_p} = \frac{\text{kinematic viscosity}}{\text{thermal diffusivity}} \qquad (2.26)$$

Since the kinematic viscosity is a measure of the molecular diffusivity of momentum, the Prandtl number is, therefore, the ratio of the molecular diffusivities of momentum and heat. It relates the temperature distribution to the velocity distribution so that, in geometrically similar systems having the same Prandtl and Reynolds numbers, the temperature distribution will be similar. When $Pr = 1$, the velocity and temperature profiles are similar, but when Pr is less than unity, the temperature gradient near a surface is less steep than the velocity gradient. The Prandtl numbers for oxygen, nitrogen and carbon dioxide lie in the range 0.7 to 0.8 at temperatures in the range 0-2800 K and are virtually independent of pressure from 1 to 10 bar. Consequently, in many combustion problems, the Prandtl number is taken to be 0.71.

Equation 2.20 is often expressed as:

$$Nu = f(Re, Pr) \qquad (2.27)$$

Correlation of experimental data

Extensive measurements have been made of heat transfer coefficients for the flow of air normal to single heated cylinders of various diameters. Since the Prandtl number for air varies very little, only the Nusselt and Reynolds numbers need to be considered. Figure 2.5 shows the results plotted on a logarithmic scale for the range of Reynolds numbers from $0.1\text{-}2 \times 10^5$. This indicates that all the data can be correlated by equations of the form:

$$Nu = B\, Re^{n} \qquad (2.28)$$

For practical purposes, the values of B and the exponent n can be taken as constant over reasonably wide ranges of Reynolds numbers as indicated in Table 2.2.

Experimental data obtained for air alone do not indicate the dependence of the Nusselt number on the Prandtl number since Pr for air is almost constant. The influence of the Prandtl number has been determined by using various fluids, including water and

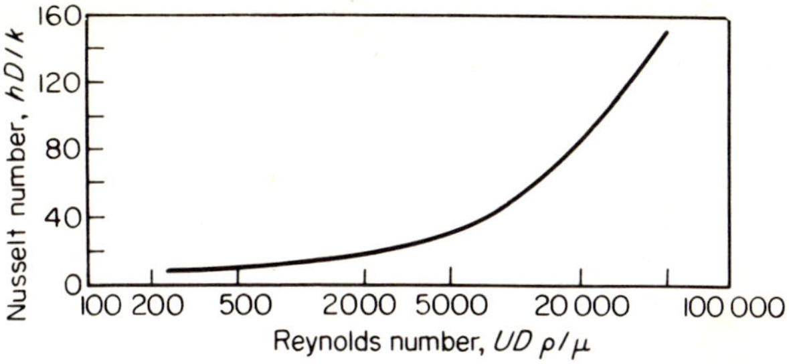

Figure 2.5 *Heat transfer data for flow of air over a 1" o.d. pipe*

Table 2.2 Values of the constants n and B for different values *Re*

Re	n	B
1-4	0.330	0.891
4-40	0.385	0.821
40-4,000	0.466	0.615
4,000-40,000	0.618	0.174
40,000-250,000	0.805	0.024

oils and it has been found convenient to combine *Nu* and *Pr* into the dimensionless quantity $Nu/Pr^{0.3}$ which, when plotted against *Re* on a logarithmic scale correlates all the experimental results on a single line as shown in Figure 2.6.

Reliable data on convective heat transfer to be found in the literature are generally confined to fairly simple cases, such as flow through tubes and ducts of various simple cross-sections, flow over flat plates and tube banks and impingement of jets on flat plates. Some examples of

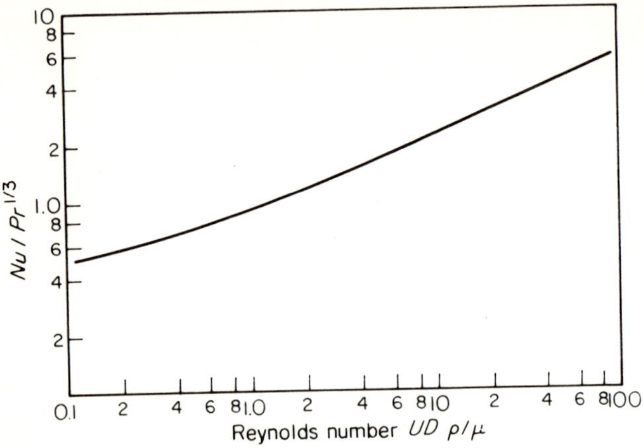

Figure 2.6 Experimental heat transfer data for various fluids flowing across cylinders of various diameters

correlations for turbulent flow are given below.

(i) *Flow inside tubes*. In laminar flow through a tube, there is no mixing of warmer and colder gases by eddy motion and the heat transfer takes place solely by conduction. Since gases have low thermal conductivities, the heat transfer coefficients are relatively small. In transitional flow, some mixing occurs by means of eddies which carry warmer gases into cooler regions, and vice versa. This mixing, even if it is only on a small scale, increases the heat transfer considerably. Figure 2.7 shows experimental values of *Nu* for air flowing through a long heated tube, plotted against *Re*. In the laminar region *Nu* increases from about 2.2 at *Re* = 200 to 5.0 at *Re* = 2100. Above this Reynolds number, *Nu* begins to increase rapidly until *Re* reaches about 8000. As *Re* is further increased, *Nu* continues to increase at a slower rate. This is because, in the fully turbulent regime, the thermal resistance of the laminar sublayer controls the heat transfer and most of the

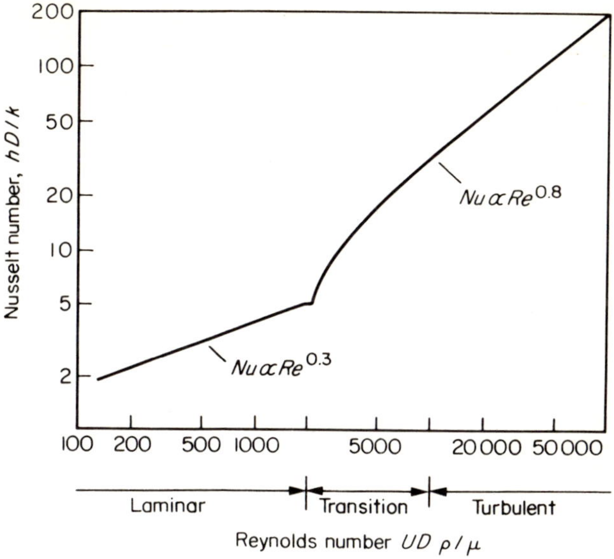

Figure 2.7 *Heat transfer data for air flowing in a pipe*

temperature drop between the bulk of the air and the surface of the tube occurs in this layer.

For a mixture of gases, the Prandtl number must be taken into account and the relationship:

$$Nu = 0.023\ Re^{0.8}\ Pr^{0.33} \tag{2.29}$$

may be used for $Re > 10{,}000$, provided that the length of the pipe is greater than 60 pipe diameters. It should be noted that the physical properties of the gas (μ, k and ρ) are evaluated at the film temperature, which is taken as the arithmetic mean of the bulk gas and surface temperatures.

(ii) *Flow over a sphere.* The definition of Nu given in Equation 2.30 may be used in the range $25 < Re < 10^5$, the properties again being evaluated at the film temperature.

$$Nu = 0.37\ Re^{0.6}\ Pr^{0.3} \tag{2.30}$$

(iii) *Tube bundles in cross flow.* Many different correlations have been derived for various arrangements of tubes, e.g. whether staggered or in line. One example for $Re > 6000$ is:

$$Nu = 0.33\ C_H\ Re^{0.6}\ Pr^{0.3} \qquad (2.31)$$

where the value of the empirical constant C_H depends on the tube arrangement and the Reynolds number. For pitch to diameter ratios between 1.25 and 1.5 the value of C_H is within 10% of unity. Formulae for other geometrical arrangements (for example, flow over plane surfaces) may be found in standard textbooks on heat transfer (Kreith, 1973).

(b) *Natural Convection*

When a temperature difference exists between a wall and a stationary fluid, temperature gradients in the fluid cause corresponding density gradients which, in turn, result in buoyancy forces and free or natural convection. Corresponding to Equation 2.20, the variables which influence the rate of heat transfer are:

$$h = f(\mu,\ \rho,\ k,\ c_p,\ \Delta T,\ \beta g,\ L) \qquad (2.32)$$

where β is the coefficient of cubical expansion and L is the characteristic dimension. Dimensional analysis shows the significance of the dimensionless modulus $\rho^2\ \beta g\ \Delta T\ L^3/\mu^2$, the Grashof number, as well as the Nusselt and Prandtl numbers. The Grashof number can be interpreted as the ratio of buoyant to viscous forces. In natural convection alone, when buoyancy forces determine the fluid velocity, the Grashof number is sufficient to determine the nature of the flow and can be used to establish dynamic similarity between two situations.

Experimental data have, therefore, been correlated by expressions of the form:

$$Nu = f(Gr,\ Pr) \qquad (2.33)$$

For a number of situations, it is possible to use

the correlation:

$$Nu = \text{constant}\ (Gr \times Pr)^m \qquad (2.34)$$

(i) *Vertical plates.* In particular, for free convection from vertical plates or cylinders the average Nusselt number over the height is:

$$Nu = 0.555\ (Gr\ Pr)^{0\cdot 25} \quad \text{for } Gr < 10^9 \qquad (2.35)$$

or

$$Nu = 0.1\ (Gr\ Pr)^{0\cdot 33} \quad \text{for } Gr > 10^9 \qquad (2.36)$$

The different constants and indices in these equations correspond to laminar (Equation 2.35) and turbulent flow (Equation 2.36).

For turbulent convection the heat transfer coefficient is given by:

$$h_c = \frac{0.1\ k\ (\rho^2\ \beta g\ \Delta T\ L^3/\mu^2)^{0\cdot 33}(c_p\ \mu/k)^{0\cdot 33}}{L} \qquad (2.37)$$

which shows that h_c is independent of the height L. Otherwise, the local heat transfer coefficient varies with height and the average value is:

$$h_c = (1/L) \int_0^L h_{cx}\ dx \qquad (2.38)$$

where h_{cx} is the local value of the coefficient at distance x from the leading edge of the solid.

(ii) *Horizontal cylinders.* McAdams gives the following correlation for the average Nusselt number from $Pr > 0.5$ and $10^3 < Gr < 10^9$:

$$Nu = 0.53\ (Gr\ Pr)^{0\cdot 25} \qquad (2.39)$$

Simpler expressions for the average heat transfer coefficients themselves are given by McAdams (1954) for horizontal cylinders with moderate surface temperatures in air:

$$\bar{h}_c = 1.32\ (\Delta T/D)^{\frac{1}{4}} \text{ for } 10^3 < Gr < 10^9 \qquad (2.40)$$

and $$\bar{h}_c = 1.25\ (\Delta T)^{1/3} \text{ for } 10^9 < Gr < 10^{12} \qquad (2.41)$$

where D is the diameter of the cylinder.

(iii) *Spheres.* Data on natural convection to or from spheres has been correlated by:

$$Nu = (h_c\ D/k) = 2 + 0.45\ (Gr\ Pr)^{0.25} \qquad (2.42)$$

For very small spheres, the Grashof number tends to zero and the Nusselt number approaches 2, which corresponds to pure conduction through a stagnant layer surrounding the sphere of thickness approximately equal to the radius.

(iv) *Horizontal square plates* (L × L m). The available data provide the following correlations:

(a) Hot plate facing up or cool plate facing down

$$Nu = (h_c\ L/k) = 0.54\ (Gr\ Pr)^{1/4} \text{ for } 10^5 < Gr < 2 \times 10^7 \qquad (2.43)$$

and

$$Nu = 0.14\ (Gr\ Pr)^{1/3} \text{ for } 2\times10^7 < Gr < 3 \times 10^{10} \qquad (2.44)$$

(b) Hot plate facing down or cool plate facing up

$$Nu = 0.27\ (Gr\ Pr)^{1/4} \text{ for } 3\times10^5 < Gr < 3 \times 10^{10} \qquad (2.45)$$

Equations 2.43, 2.44 and 2.45 may be used for horizontal circular discs of diameter D if L is replaced by $0.9D$, and for rectangular plates if L is set equal to the average side length.

2.5 SOLUTION OF THE BOUNDARY LAYER EQUATIONS

In the preceding section, dimensional analysis has been used to determine dimensionless groups for correlating experimental data. In addition to this technique, analytical methods of approach may be used, involving mathematical solution of the equations for the hydrodynamic and thermal boundary layers.

2.5.1 Exact Boundary Layer Analysis

Essentially, exact mathematical analysis involves the simultaneous solution of the differential equations for momentum conservation and continuity, applying the appropriate boundary conditions to obtain expressions for the velocity distribution and thickness of the hydrodynamic boundary layer. The solution of these was first obtained numerically by Blasius for the laminar boundary layer on a flat surface. In order to evaluate the rate of convective heat transfer, this velocity distribution is used, together with the differential equation of energy conservation, to determine the temperature distribution in the thermal boundary layer.

Exact mathematical solutions have only been obtained for certain cases of laminar flow in simple systems, such as flat surfaces, where there is no change in pressure gradient along the surface.

2.5.2 Approximate Boundary Layer Analysis

The momentum conservation equation is a partial differential equation with variable coefficients and its solution is difficult to obtain even for the simple geometry of a flat surface. The mathematical difficulties of an exact solution may be avoided by an approximate analysis which is based on the von Karman integral momentum equation for the boundary layer, together with an assumed velocity distribution (described by a cubic equation, for example). The temperature distribution in the thermal boundary layer is often assumed to be of the same form and this may be used with the integral energy equation to obtain the required heat transfer relation. It has been found that even a very rough assumption for the velocity distribution will yield satisfactory results and the method has been applied to many geometrical arrangements which cannot be treated by exact mathematical analysis. The technique is not limited to laminar flow, but can also be applied to certain simple cases of turbulent flow.

2.6 THE ANALOGY BETWEEN HEAT AND MASS TRANSFER

In design applications where no relevant practical heat transfer data is available for the geometrical and flow configuration under consideration, it is sometimes possible to evaluate heat transfer coefficients by employing various model techniques. The most reliable of these are based on the analogy between heat and mass transfer under conditions of dynamic similarity, the heat transfer surfaces of the appliance being replaced by mass transfer surfaces in the model. The heat and mass transfer processes are related by the Chilton-Colburn analogy (Chilton and Colburn, 1934) which states that, under conditions of geometric, kinematic and dynamic similarity, the dimensionless heat and mass transfer functions, j, are equal where:

$$\text{for heat transfer} \quad j = \frac{h_c}{G\, c_p} Pr^{2/3} \tag{2.32}$$

$$\text{for mass transfer} \quad j = \frac{h_D \rho}{G} Sc^{2/3} \tag{2.33}$$

where h_D is the convective mass transfer coefficient, the Schmidt number (Sc) is $\mu/\rho D_{ab}$ and D_{ab} is the diffusivity of a in b.

2.6.1 Similarity criteria in models

Geometrical similarity, the simplest of the model laws, implies that every linear dimension of the model bears the same ratio to the corresponding dimension of the full scale appliance. Although this can be achieved for isothermal systems, some distortion is necessary when sources of heat release, such as flames, are involved.

Kinematic similarity implies that the fluid follows geometrically similar paths in corresponding intervals of time. This means that there is a velocity scale ratio which is maintained constant between corresponding points of the model and the prototype.

Dynamic similarity requires that the force ratios, causing acceleration of masses in the corresponding systems are maintained constant. This condition is usually met by ensuring that the Reynolds numbers are equal.

2.6.2 Measurement of Convective Mass Transfer Coefficients

Two techniques for measuring mass transfer coefficients in models have been applied by Francis *et al.*(1967) in their study of convection in rapid heating furnaces, naphthalene sublimation and electrolytic techniques.

Naphthalene sublimation

Measurement of mass transfer coefficients by the sublimation of naphthalene in an air stream was first used by Winding and Cheney (1948) in their work on heat transfer to a bank of tubes. The rate of sublimation from a surface may be determined in two ways; either by measurement of the loss of weight from the surface or by measurement of the change in profile of the surface. The former method will only give overall values for the mass transfer rates from the whole of the naphthalene surface, whereas the latter method enables local as well as overall values to be measured.

The surface in the model corresponding to the heat transfer surfaces in the appliance are coated with naphthalene, either by casting in a polished brass mould or painting naphthalene on a copper former. Air at a carefully controlled temperature is passed through the model at a constant flow rate to simulate the flow of hot gases. The resulting weight loss or dimension change of the naphthalene coated surfaces is measured.

Electrolytic techniques

Electrolytic techniques for determining mass transfer coefficients are based on measuring the limiting, diffusion-controlled current to an electrode in a specially designed electrolytic cell.

In the technique adopted by Francis, nickel electrodes were formed by electroplating the surfaces of a model corresponding to the heat transfer surfaces.

3 Convective Heat Transfer from Flames

In some conventional combustion appliances, convective heat transfer is comparatively unimportant and the performance can often be predicted adequately by using a very approximate coefficient, such as that suggested by Trinks and Mawhinney (1961). In plant where the heat transfer surface is sufficiently removed from the flame and the combustion products are at a relatively low temperature, data may be obtained from existing correlations given in the literature (for example, for flow through or over tubes of various cross-sections). In these cases, the general practice is to assume that the mixture of gases constituting the combustion products possess the same transport properties as air.

The greatest difficulties arise in calculations of the convective heat transfer to surfaces exposed to high temperature gases, especially when the surface is cold. Three modifications to the calculation may be required to allow for the following: (1) that the transport properties of the combustion products can no longer be regarded as identical with those of air; (2) that the use of the arithmetic mean of the surface and gas temperatures as the appropriate film temperature is probably inadequate when there are large variations in physical properties with temperature, and (3) that chemical recombination reactions occur when the hot gases are cooled and, as a consequence, there may be some heat release within the boundary layer.

3.1 THERMAL AND TRANSPORT PROPERTIES OF GAS MIXTURES

A prerequisite to the calculation of the properties of gaseous mixtures is a knowledge of the composition of the mixture. A method for determining the equilibrium composition of combustion products is given in Appendix 1.

3.1.1 Specific Heat

The specific heat (per kilogramme) of a flame gas mixture of n components at various temperatures may be evaluated from the expression:

$$c_{p_{mix}} = \frac{10^3}{M_e} \sum_{i=1}^{n} C_{p_i} \, x_i \, M_i \qquad (3.1)$$

where C_{p_i} = the molar heat capacity of the individual component i

x_i = the mole fraction of component i

M_i = the molecular weight of component i

M_e = the mean molecular weight of the gas mixture.

Values of C_{p_i} for the most common components of combustion products are given in Tables 3.1 to 3.6.

3.1.2 Viscosity

The viscosities of gaseous mixtures, unlike specific heats, cannot be calculated by a simple mixing rule such as that used in Equation 3.1. Intermolecular forces must be taken into account and although it is possible to use the kinetic theory rigorously, the complexity of the procedure renders it unsuitable for practical use. Simpler expressions, involving a limited number of approximations, have been derived by Wilke (1950) and Brokaw (1958) and the calculated values from these formulae agree closely with the estimates obtained from theory.

Table 3.1. Molar heat capacities, viscosities and thermal conductivities of carbon dioxide (molecular weight = 44)

Temperature K	C_p J mol^{-1} K^{-1}	μ N s m^{-2}	k W m^{-1} K^{-1}
300	37.2	15.2×10^{-6}	18.2×10^{-3}
400	41.4	19.6×10^{-6}	25.9×10^{-3}
600	47.4	27.1×10^{-6}	40.7×10^{-3}
800	51.5	33.6×10^{-6}	54.5×10^{-3}
1000	54.3	39.3×10^{-6}	67.2×10^{-3}
1200	56.4	44.5×10^{-6}	78.9×10^{-3}
1400	57.8	49.5×10^{-6}	89.7×10^{-3}
1600	58.9	54.0×10^{-6}	99.9×10^{-3}
1800	59.7	58.4×10^{-6}	109.3×10^{-3}
2000	60.4	62.5×10^{-6}	118.3×10^{-3}
2200	60.9	66.5×10^{-6}	126.8×10^{-3}
2400	61.3	70.3×10^{-6}	134.9×10^{-3}
2600	61.7	74.0×10^{-6}	142.8×10^{-3}
2800	62.0	77.6×10^{-6}	150.5×10^{-3}
3000	62.3	81.2×10^{-6}	158.1×10^{-3}

Table 3.2. Molar heat capacities, viscosities and thermal conductivities of carbon monoxide (molecular weight = 28)

Temperature K	C_p J mol^{-1} K^{-1}	μ N s m^{-2}	k W m^{-1} K^{-1}
300	29.2	17.8×10^{-6}	26.7×10^{-3}
400	29.4	21.9×10^{-6}	33.2×10^{-3}
600	30.5	29.0×10^{-6}	45.5×10^{-3}
800	31.9	35.1×10^{-6}	57.5×10^{-3}
1000	33.2	40.6×10^{-6}	68.9×10^{-3}
1200	34.2	45.6×10^{-6}	79.6×10^{-3}
1400	34.9	50.4×10^{-6}	89.8×10^{-3}
1600	35.4	55.1×10^{-6}	99.5×10^{-3}
1800	35.9	59.5×10^{-6}	108.6×10^{-3}
2000	36.3	63.7×10^{-6}	117.3×10^{-3}
2200	36.6	67.8×10^{-6}	125.7×10^{-3}
2400	36.8	71.7×10^{-6}	133.7×10^{-3}
2600	36.9	75.5×10^{-6}	141.5×10^{-3}
2800	37.1	78.2×10^{-6}	149.0×10^{-3}
3000	37.2	82.8×10^{-6}	156.2×10^{-3}

Table 3.3. Molar heat capacities, viscosities and thermal conductivities of water vapour (molecular weight = 18)

Temperature K	C_p J mol^{-1} K^{-1}	μ N s m^{-2}	k W m^{-1} K^{-1}
300	33.6	10.2×10^{-6}	27.2×10^{-3}
400	34.3	13.9×10^{-6}	37.6×10^{-3}
600	36.3	21.2×10^{-6}	60.7×10^{-3}
800	38.7	28.5×10^{-6}	86.7×10^{-3}
1000	41.2	35.5×10^{-6}	114.6×10^{-3}
1200	43.7	42.1×10^{-6}	143.6×10^{-3}
1400	46.0	48.4×10^{-6}	173.3×10^{-3}
1600	48.0	54.5×10^{-6}	203.3×10^{-3}
1800	49.7	60.3×10^{-6}	232.5×10^{-3}
2000	51.1	65.8×10^{-6}	260.8×10^{-3}
2200	52.4	70.8×10^{-6}	287.1×10^{-3}
2400	53.4	75.9×10^{-6}	313.6×10^{-3}
2600	54.3	80.8×10^{-6}	339.3×10^{-3}
2800	55.0	85.4×10^{-6}	363.6×10^{-3}
3000	55.7	90.0×10^{-6}	387.9×10^{-3}

Table 3.4. Molar heat capacities, viscosities and thermal conductivities of hydrogen (molecular weight = 2)

Temperature K	C_p J mol^{-1} K^{-1}	μ N s m^{-2}	k W m^{-1} K^{-1}
300	28.9	8.9×10^{-6}	-
400	29.2	10.9×10^{-6}	-
600	29.3	14.2×10^{-6}	31.3×10^{-3}
800	29.7	17.1×10^{-6}	36.3×10^{-3}
1000	30.2	19.8×10^{-6}	42.8×10^{-3}
1200	30.9	22.3×10^{-6}	49.3×10^{-3}
1400	31.8	24.6×10^{-6}	55.9×10^{-3}
1600	32.8	26.9×10^{-6}	62.5×10^{-3}
1800	33.6	29.0×10^{-6}	68.9×10^{-3}
2000	34.3	31.0×10^{-6}	75.2×10^{-3}
2200	35.0	33.0×10^{-6}	81.4×10^{-3}
2400	35.6	34.9×10^{-6}	87.4×10^{-3}
2600	36.2	36.7×10^{-6}	93.2×10^{-3}
2800	36.7	38.5×10^{-6}	98.9×10^{-3}
3000	37.1	40.3×10^{-6}	104.5×10^{-3}

Table 3.5. Molar heat capacities, viscosities and thermal conductivities of nitrogen (molecular weight = 28)

Temperature	C_p	μ	k
K	$J\ mol^{-1}\ K^{-1}$	$N\ s\ m^{-2}$	$W\ m^{-1}\ K^{-1}$
300	29.1	17.8×10^{-6}	26.8×10^{-3}
400	29.3	21.7×10^{-6}	32.9×10^{-3}
600	30.1	28.5×10^{-6}	44.3×10^{-3}
800	31.5	34.4×10^{-6}	55.6×10^{-3}
1000	32.7	39.7×10^{-6}	66.5×10^{-3}
1200	33.7	44.8×10^{-6}	77.2×10^{-3}
1400	34.5	49.5×10^{-6}	87.2×10^{-3}
1600	35.2	54.0×10^{-6}	96.6×10^{-3}
1800	35.6	58.3×10^{-6}	105.6×10^{-3}
2000	36.0	62.4×10^{-6}	114.1×10^{-3}
2200	36.3	66.4×10^{-6}	122.3×10^{-3}
2400	36.6	70.2×10^{-6}	130.2×10^{-3}
2600	36.8	73.9×10^{-6}	137.7×10^{-3}
2800	36.9	77.6×10^{-6}	145.1×10^{-3}
3000	37.1	81.1×10^{-6}	152.2×10^{-3}

Table 3.6. Molar heat capacities, viscosities and thermal conductivities of oxygen (molecular weight = 32)

Temperature K	C_p J mol^{-1} K^{-1}	μ N s m^{-2}	k W m^{-1} K^{-1}
300	29.4	20.6×10^{-6}	27.5×10^{-3}
400	30.1	25.7×10^{-6}	34.9×10^{-3}
600	32.1	34.1×10^{-6}	49.2×10^{-3}
800	33.8	41.5×10^{-6}	62.6×10^{-3}
1000	34.9	48.1×10^{-6}	74.8×10^{-3}
1200	35.7	54.1×10^{-6}	85.9×10^{-3}
1400	36.3	59.7×10^{-6}	96.3×10^{-3}
1600	36.8	65.1×10^{-6}	106.5×10^{-3}
1800	37.3	70.4×10^{-6}	116.5×10^{-3}
2000	37.8	75.4×10^{-6}	126.3×10^{-3}
2200	38.3	80.2×10^{-6}	135.9×10^{-3}
2400	38.7	84.9×10^{-6}	145.3×10^{-3}
2600	39.2	89.4×10^{-6}	154.6×10^{-3}
2800	39.6	93.8×10^{-6}	163.8×10^{-3}
3000	40.0	98.1×10^{-6}	172.8×10^{-3}

The expressions given by Wilke are:

$$\mu_{mix} = \sum_{i=1}^{n} \frac{\mu_i}{1 + \sum\limits_{\substack{j=1 \\ j \neq i}}^{n} \phi_{ij} \frac{x_j}{x_i}} \qquad (3.2)$$

where μ_i and μ_j are the viscosities of pure components i and j,

and

$$\phi_{ij} = \frac{\left[1 + \left(\frac{\mu_i}{\mu_j}\right)^{\frac{1}{2}} \left(\frac{M_j}{M_i}\right)^{\frac{1}{4}}\right]^2}{2 \times 2^{\frac{1}{2}} \left[1 + \frac{M_i}{M_j}\right]^{\frac{1}{2}}} \qquad (3.3)$$

Equation 3.3 is derived from the assumption that all molecules behave as rigid spheres.

The summation for μ_{mix} is of the form:

$$\mu_{mix} = \frac{\mu_1}{1 + (x_2/x_1)\phi_{12} + (x_3/x_1)\phi_{13} + \ldots} + \frac{\mu_2}{1 + (x_1/x_2)\phi_{21} + (x_3/x_2)\phi_{23} + \ldots} + \frac{\mu_3}{1 + (x_1/x_3)\phi_{31} + (x_2/x_3)\phi_{32} + \ldots} + \qquad (3.4)$$

Individual values of μ_i are given in Tables 3.1 to 3.6.

3.1.3 Thermal Conductivity

A similar type of mixing rule for the thermal conductivity of gas mixtures has been derived by

Mason and Saxena (1958),

$$k_{mix} = \sum_{i=1}^{n} \frac{x_i\, k_i}{\sum\limits_{\substack{j=1 \\ j \neq 1}}^{n} x_j\, \phi_{ij}} \qquad (3.5)$$

ϕ_{ij} may be taken as having the same value as that given by Equation 3.3. Thermal conductivities of constituent gases are given in Tables 3.1 to 3.6. Examples of the results of such calculations which may prove of assistance in heat transfer calculations are given in Figures 3.1, 3.2 and 3.3, in which the specific heats, viscosities, and thermal conductivities for the combustion products of stoichiometric methane/air mixtures are plotted as a function of temperature.

3.2 REFERENCE TEMPERATURE

There has been a considerable amount of confusion in the literature concerning the dependence of convective heat transfer coefficients on temperature in the presence of large temperature differences. Many different empirical schemes have been suggested, but the principal choice appears to be between evaluation of all the properties of the combustion products at either the bulk gas temperature or the mean film temperature. The mean film temperature is the most logical, but care must be taken in applying existing correlations from the field of chemical engineering, since some of these correlations have been based on bulk temperature.

Some difficulties arise in the case of high temperature flames, due to the large and non-linear variation of thermophysical properties with temperature. Film values in such systems are best represented by weighted averages of the particular property between the wall (surface) temperature (T_w) and the free stream gas temperature (T_e). The weighted average viscosity ($\overline{\mu}$) is defined by,

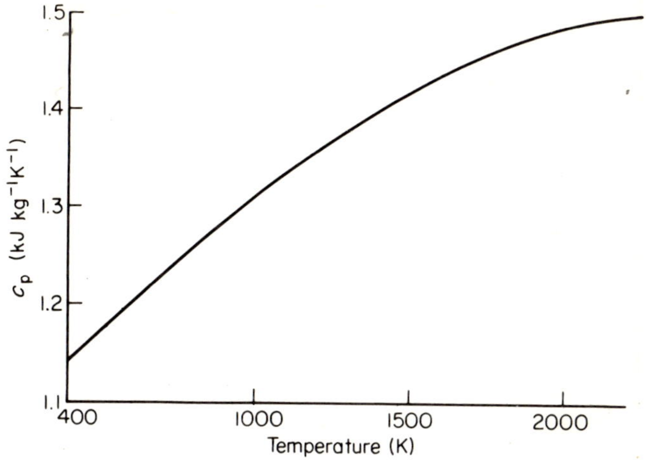

Figure 3.1 *Specific heat of stoichiometric methane/air combustion products*

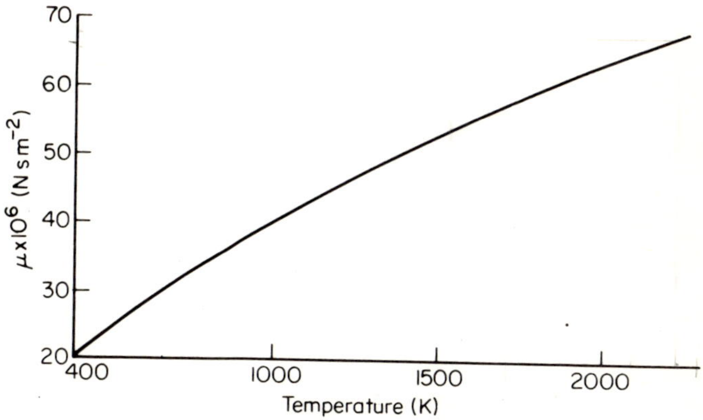

Figure 3.2 *Viscosity of stoichiometric methane/air combustion products*

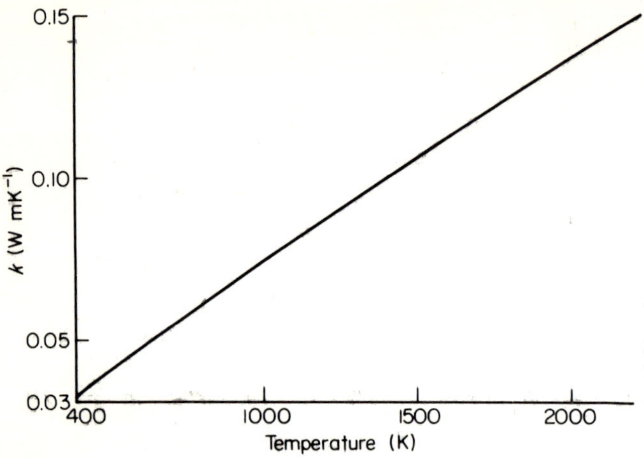

Figure 3.3 Thermal conductivity of stoichiometric methane/air combustion products

$$\bar{\mu} = \frac{\int_{T_w}^{T_e} \mu \, dT}{(T_e - T_w)} \tag{3.6}$$

Weighted averages for thermal conductivities, specific heats and Prandtl numbers are similarly defined.

3.3 CHEMICAL RECOMBINATION REACTIONS

At the high temperatures of some flames, the primary combustion products dissociate into more unstable species, such as free radicals and atoms, and these dissociation reactions involve absorption of energy.

Examples of such dissociation equilibria in flames containing C, H and O are:

$$H_2O \rightleftharpoons \tfrac{1}{2} H_2 + OH$$

$$H_2 \rightleftharpoons 2 H$$

$$H_2O \rightleftharpoons H_2 + \tfrac{1}{2} O_2$$

$$CO_2 + H_2 \rightleftharpoons CO + H_2O$$

Table 3.7. Composition of stoichiometric methane/air combustion products

	Mole fractions				
Components	1400 K	1600 K	1800 K	2000 K	2222 K
N_2	0.715	0.715	0.714	0.713	0.709
H_2O	0.190	0.190	0.189	0.188	0.184
CO_2	0.095	0.095	0.0941	0.0918	0.0854
CO	0	0.00016	0.00088	0.00307	0.0090
O_2	0	0.00014	0.00052	0.00168	0.0047
H_2	0	0.00011	0.00045	0.00135	0.00354
OH	0	0.00003	0.00018	0.00072	0.00244
NO	0	0.00005	0.00021	0.00069	0.00191
O	0	0	0	0.00003	0.00021
H	0	0	0.00001	0.00006	0.00040

The effect of dissociation reactions on the composition of the combustion products of stoichiometric methane/air flames is illustrated in Table 3.7. These data show that, without dissociation, the combustion products should contain 19.0 mole per cent of H_2O and 9.5 mole per cent of CO_2. At the adiabatic flame temperature of 2222 K, however, the products contain only 18.4% of H_2O and 8.54% of CO_2 together with small amounts of CO, O_2, H_2, OH, NO, O and H.

Table 3.8. Composition of stoichiometric methane/oxygen combustion products

Component	Mole fractions				
	1400 K	1800 K	2200 K	2600 K	3050 K
H_2O	0.667	0.664	0.647	0.578	0.391
CO_2	0.333	0.331	0.310	0.239	0.111
CO	0	0.00192	0.0182	0.0728	0.156
O_2	0	0.00135	0.0113	0.0423	0.0847
H_2	0	0.00099	0.0071	0.0279	0.0753
OH	0	0.00042	0.0053	0.0283	0.0847
O	0	0	0.00031	0.0051	0.0472
H	0	0.00001	0.00048	0.0064	0.0497

If oxygen is used instead of air, the effect of dissociation is much more evident as shown in Table 3.8. At the adiabatic flame temperature of 3050 K, the combustion products of a stoichiometric methane/oxygen flame contain 15.6% CO, 8.47% O_2, 7.53% H_2, 8.47% OH radicals, 4.72% O atoms and 4.97% H atoms.

It is reasonable to assume that when the combustion products are cooled, these unstable species will recombine. Provided that sufficient time is available for the recombination reactions to take place, the composition of the gas mixture at any temperature will attain its equilibrium value. Normally, recombination reactions of free atoms and radicals are very rapid in the presence of a third body (a solid surface or other larger molecules). The third body is necessary to abstract the considerable amount of heat that is released during the reactions. Some examples of typical recombination reactions and molecular equilibria, together with the amount of heat release, are:

H+H+third body	→	H_2+third body	ΔH = -436.3 kJ mol^{-1}
$\frac{1}{2}$ H_2+OH	→	H_2O	ΔH = -281 kJ mol^{-1}
CO+H_2O	→	CO_2+H_2	ΔH = - 41.2 kJ mol^{-1}

The negative sign for ΔH, according to the accepted thermochemical convention, indicates that heat is released, that is, the reaction is exothermic. The mechanism of heat transfer from partially dissociated gas streams to a cold surface may be considered qualitatively in terms of the conditions existing in the boundary layer surrounding the cold body. Outside the boundary layer, that is, in the external flow, any free radicals and atoms are unaffected by the presence of the solid and hence their concentration remains constant. However, inside the boundary layer a temperature gradient exists and, in order to maintain equilibrium, some of the atoms and radicals recombine, with a consequent reduction in their concentrations. This difference in concentration inside and outside the boundary layer provides a driving force for the atoms and radicals to diffuse towards the region of low concentration where they recombine exothermically. Thus, when heat is transferred to surfaces at lower temperatures, chemical energy released on the recombination of the dissociated species augments the normal mode of heat transfer by conduction through the boundary layer. This additional mode of heat transfer depends on the energy of dissociation and the velocities of diffusion of the individual reactive species.

Before attempting to assess the effects of variation of thermal and transport properties on rates of heat transfer, it is necessary to define the conditions at the gas/solid interface if existing techniques of predicting heat transfer rates are to be adapted to high temperature conditions.

In the simpler analysis of heat transfer from reacting systems, it is customary to consider two, physically distinct, cases. Firstly, the local conditions in the boundary layer may be such that the exothermic recombination reactions of the dissociated species are very slow compared with their residence times near the surface. The flow is said to be chemically frozen and no effect on heat transfer is to be expected, provided that the surface is noncatalytic to the recombination reactions. If, at the other extreme, conditions are such that the reactions are rapid enough for the gas composition to maintain local chemical equilibrium everywhere within the boundary layer, an increase in the rates of heat transfer will occur as a result of the heat released by recombination reactions. Intermediate conditions are obviously possible, but these are difficult to compute in the absence of relevant data on reaction rates and a knowledge of the catalytic properties of surfaces for the reactions involved. On the basis that the wall is fully catalytic, both Lees (1956) and Rosner (1961) have predicted that the heat transfer for given free stream conditions is independent of the boundary layer flow being frozen or equilibrium. It is important to define the distinction between the frozen and equilibrium properties of a dissociated gas mixture.

The frozen specific heat and frozen thermal conductivity are computed from the normal properties of the constituents of the gas mixture as described in Section 3.1, whereas the equilibrium specific heat and equilibrium thermal conductivity take into account the enthalpy involved in the dissociation-recombination reactions. It is these latter physical properties which are effective in reacting gas mixtures in the equilibrium boundary layer case. Since viscosity is not a thermal property, its value at a given temperature is independent of the flow being frozen or equilibrium, provided that the correct gas mixture composition is used in its evaluation.

3.3.1 Enthalpy

The term 'enthalpy' is normally regarded as being the sensible heat content of the gas and consequently the enthalpy difference between temperatures T_1 and T_2 is defined as:

$$\Delta H = (H_{T_1} - H_{T_2}) = \int_{T_1}^{T_2} C_p \, dT \qquad (3.7)$$

Since C_p varies with temperature, this is not a very suitable method of evaluation and it has become customary to tabulate values of enthalpy at various temperatures as $(H_T - H_{298})$, based on the convention that H_{298} is zero, that is 298 K is a reference temperature. A convenient source of such data is the Joint Army, Navy and Air Force (J.A.N.A.F) Thermochemical Tables (1965) which give values of $(H - H_{298})$ for all gaseous components at 100 K temperature intervals. The sensible enthalpy per mole of combustion products may be calculated from:

$$H = \sum_{i=1}^{n} x_i \, (H - H_{298})_i \qquad (3.8)$$

A useful concept when considering thermal effects in reacting gases is that of 'total enthalpy' which, for a single component, may be defined as the sum of the sensible enthalpy and the standard heat of formation.

For a multicomponent gas mixture:

$$H_{total} = \sum_{i=1}^{n} x_i \, (H - H_{298})_i + \sum_{i=1}^{n} x_i \, (\Delta H_f^0)_i \qquad (3.9)$$

where $(\Delta H_f^0)_i$ is the standard heat of formation of component i at the reference temperature.

The advantage of using total enthalpy is that, for a mixture of dissociated gases which recombine at lower temperatures, the total enthalpy difference between two temperatures includes the heat of the recombination reactions in addition to the sensible enthalpy. This follows from the standard thermochemical method of calculating heats of reaction:

$\Delta H^0_{\text{reaction}}$ = sum of heats of formation of products - sum of heats of formation of reactants

It must be emphasised that the correct equilibrium composition of the gas mixture must be used at the temperatures concerned, that is, the mole fractions of the components are different at different temperatures.

In the following discussion, total enthalpy will be designated by H and total enthalpy difference by ΔH. Sensible enthalpy will be given the symbol H_s.

In the context of convective heat transfer, the thermal properties per unit mass (kg) of combustion products are more significant than are the molar properties. This is an important distinction since the molecular weight of a reacting gas mixture varies with temperature. The enthalpy per unit mass (h) is obtained by dividing the molar enthalpy (H) by the mean molecular weight of the combustion products (in kg) at that temperature.

3.3.2 Equilibrium Specific Heats

By analogy with the normal (frozen) specific heat, the equilibrium specific heat may be defined as equal to the gradient of the total enthalpy versus temperature curve, that is,

$$c_{p_{eq}} = \frac{dh}{dT} \qquad (3.10)$$

The weighted average specific heat is then:

$$\bar{c}_{p_{eq}} = \frac{\int_{T_w}^{T_e} c_{p_{eq}} \, dT}{T_e - T_w} = \frac{\Delta h}{\Delta T} \qquad (3.11)$$

The total enthalpy difference (Δh) between two temperatures can be expressed as the sum of the sensible enthalpy difference (Δh_s) and the dissociation enthalpy difference (Δh_{chem}).

The frozen specific heat may, therefore, be written as:

$$\bar{c}_{p_f} = (\Delta h - \Delta h_{chem})/\Delta T \qquad (3.12)$$

The frozen and equilibrium specific heats of stoichiometric methane/air and methane/oxygen combustion products are given in Tables 3.9 and 3.10 respectively.

Table 3.9. Frozen and equilibrium specific heats of stoichiometric methane/air combustion products

Temperature K	c_{p_f} (kJ kg^{-1} K^{-1})	$c_{p_{eq}}$ (kJ kg^{-1} K^{-1})
400	1.15	1.15
600	1.19	1.19
800	1.26	1.26
1000	1.34	1.31
1200	1.37	1.37
1400	1.40	1.40
1600	1.43	1.47
1800	1.46	1.57
2000	1.48	1.76
2222	1.50	2.18

Tables 3.9 and 3.10 clearly indicate that the values of the frozen and equilibrium specific heats are identical up to 1400 K, that is, in the region where there is no dissociation of the combustion

Table 3.10. Frozen and equilibrium specific heats of stoichiometric methane/oxygen combustion products

Temperature K	c_{p_f}	$c_{p_{eq}}$
	kJ kg^{-1} K^{-1}	
400	1.37	1.37
600	1.50	1.50
800	1.61	1.61
1000	1.71	1.71
1200	1.80	1.80
1400	1.88	1.88
1600	1.94	2.00
1800	1.99	2.21
2000	2.04	2.65
2200	2.07	3.46
2400	2.10	4.78
2600	2.12	6.71
2800	2.14	9.29
3000	2.15	12.39

products. At higher temperatures, the equilibrium specific heat begins to increase more rapidly than the frozen value, being 45% greater at the flame temperature of the stoichiometric methane/air flame. At the flame temperature of methane/oxygen mixtures (3050 K) the equilibrium specific heat is 5.8 times the frozen value.

3.3.3 Equilibrium Thermal Conductivity

Equilibrium thermal conductivity (k_{eq}) includes both the contribution of frozen conductivity and the reacting conductivity due to the diffusion and exothermic recombination of dissociated species. The various reactive species diffuse towards the cold surface at different velocities and allowance for this effect is often necessary. On a quantitative basis, a dimensionless group known as the Lewis number is used. This is equal to the ratio of the diffusion coefficient of the species through the gas mixture ($D_{i\text{-mix}}$) to the thermal diffusivity of that mixture ($k/\rho\, c_p$):

$$Le_i = \frac{D_{i\text{-mix}}}{k/\rho\, c_p} \qquad (3.13)$$

The Lewis number is thus an indication of the rate at which heat can be transported through a mixture by diffusion in comparison with conduction.

The process of diffusion and exothermic recombination of reactive species may be regarded as increasing the effective thermal conductivity of the gaseous products above the frozen value which applies in the absence of chemical reaction. Hirschfelder (1957) has shown that the equilibrium thermal conductivity can be related to the frozen thermal conductivity, equilibrium and frozen specific heats by:

$$k_{eq} = k_f \left[1 + Le \left(\frac{c_{p_{eq}}}{c_{p_f}} - 1\right)\right] \qquad (3.14)$$

This relationship only applies strictly to simple equilibria of the type $A = B$, but it can be generalised to multicomponent systems with reasonable accuracy.

If it is assumed that heat transfer by diffusion and recombination takes place at the same rate as by conduction, then the Lewis number will be equal to unity and Equation 3.14 is simplified to:

$$k_{eq} = k_f \; c_{p_{eq}} \; c_{p_f} \qquad (3.15)$$

This approximation is sufficiently accurate for flames with air as oxidant when the temperature and concentration of hydrogen atoms are relatively low.

3.4 MODIFICATION OF EMPIRICAL HEAT TRANSFER CORRELATIONS

It was shown in Chapter 2 that heat transfer data can be correlated by expressions of the form:

$$Nu = A\ Pr^{a}\ Re^{b} \tag{3.16}$$

where A, a and b are determined from plots of experimental results. As an alternative to the Nusselt number, it is sometimes convenient to use the Stanton number, St, where:

$$St = h_c / G\ c_p$$

The Stanton and Nusselt numbers are related by the expression:

$$St = Nu / Re\ Pr \tag{3.17}$$

Expressing the empirical relationships in Stanton number form:

$$St = h_c / G\ c_p = B\ Pr^{x}\ Re^{y} \tag{3.18}$$

When considering convective heat transfer from reacting gas mixtures, it is more convenient to calculate the heat flux, that is, the heat transfer per unit area of surface under the prevailing conditions, rather than the heat transfer coefficient.

$$\text{Heat flux} \quad q = h_c\ \Delta T \tag{3.19}$$

Thus, from Equation 3.18:

$$q = B\ Pr^{x}\ Re^{y}\ G\ c_p\ \Delta T \tag{3.20}$$

In the equilibrium boundary layer case, the frozen specific heat and frozen Prandtl number must be replaced by their equilibrium values. The Reynolds number is unchanged, since it does not involve any thermophysical properties.

Hence,

$$q_{eq} = B\ Pr_{eq}^{x}\ Re^{y}\ G\ c_{p_{eq}}\ \Delta T \qquad (3.21)$$

Since, $Pr = c_p\ \mu/k$:

$$\frac{Pr_{eq}}{Pr_f} = \frac{c_{p_{eq}}\ \mu/k_{eq}}{c_{p_{eq}}\ \mu/k_f} \qquad (3.22)$$

or,

$$Pr_{eq} = Pr_f\ \left(\frac{c_{p_{eq}}}{c_{p_f}}\right)\left(\frac{k_f}{k_{eq}}\right) \qquad (3.23)$$

Substituting for $c_{p_{eq}}$, c_{p_f} and k_f/K_{eq} from Equations 3.11, 3.12 and 3.14:

$$Pr_{eq} = Pr_f \left[1 + (Le - 1)\ \frac{\Delta h_{chem}}{\Delta h}\right]^{-1} \qquad (3.24)$$

Substituting for Pr_{eq} from Equation 3.24 and $c_{p_{eq}}\ \Delta T$ from Equation 3.11, Equation 3.21 becomes:

$$q_{eq} = B\ Pr_f^{x}\ Re^{y} \left[1 + (Le - 1)\ \frac{\Delta h_{chem}}{\Delta h}\right]^{-x} G\ \Delta h \qquad (3.25)$$

If the Lewis number correction is ignored (that is $Le = 1$) then Equation 3.25 reduces to:

$$q_{eq} = B\ Pr^{x}\ Re^{y}\ G\ \Delta h \qquad (3.26)$$

Comparing Equation 3.26 with its counterpart for low temperature conditions (Equation 3.20), it is apparent that even if fast diffusion effects are not taken into account, the *total* enthalpy difference (Δh) is a much more realistic driving force for heat transfer than is temperature difference.

In correcting heat flux for Lewis number effects, it is usually adequate to consider only the diffusion and recombination of hydrogen atoms since, although heat release by the recombination of heavier atoms and radicals such as O and OH will be comparable to that of H atoms, it will be limited by their lower diffusion coefficients which lead to Lewis numbers near unity. It must be appreciated, however, that the term (Δh) is the total enthalpy difference across the boundary layer and includes the heat of all recombination reactions. This implies that the Lewis numbers of all species apart from hydrogen atoms have been assumed to be unity. Since the Lewis number correction factor is confined to hydrogen atoms, the term Δh_{chem} in Equation 3.25 is evaluated from:

$$\Delta h_{chem} = \alpha_H \, (\Delta h^0_f)_H$$

where α_H is the mass fraction of hydrogen atoms in the free stream and $(\Delta h^0_f)_H$ is the heat of hydrogen atom recombination (216.4 MJ kg^{-1}). At the flame temperatures of fuel-air flames the concentration of hydrogen atoms is low enough to make the ratio $\Delta H_{chem}/\Delta h$, and hence the Lewis number augmentation factor $\left[1 + (Le - 1)\frac{\Delta h_{chem}}{\Delta h}\right]^{-x}$, negligible. In these cases correlations of the type of Equation 3.26 may be used.

3.4.1 Calculation of the Lewis Number

In order to calculate the Lewis number for hydrogen atoms diffusing through combustion products, a knowledge of the diffusion coefficients is necessary.

Binary diffusion coefficients for H atoms into other gases may be evaluated from the kinetic theory of gases, as developed by Hirschfelder *et al.* (1954). The results of such calculations are given in Table 3.11.

Multicomponent diffusion coefficients are obtained by combining the binary coefficients using an

Table 3.11. Binary diffusion coefficients for the diffusion of H atoms through gases

Temperature K	D_{H-CO_2}	D_{H-CO}	D_{H-H_2O} $m^2\ s^{-1}$	D_{H-H_2}	D_{H-N_2}	D_{H-O_2}
1600	1.64×10^{-3}	1.90×10^{-3}	2.34×10^{-3}	3.17×10^{-3}	1.88×10^{-3}	2.01×10^{-3}
1800	2.00×10^{-3}	2.32×10^{-3}	2.84×10^{-3}	3.85×10^{-3}	2.29×10^{-3}	2.45×10^{-3}
2000	2.38×10^{-3}	2.76×10^{-3}	3.38×10^{-3}	4.58×10^{-3}	2.72×10^{-3}	2.92×10^{-3}
2200	2.80×10^{-3}	3.23×10^{-3}	3.95×10^{-3}	5.36×10^{-3}	3.18×10^{-3}	3.42×10^{-3}
2400	3.23×10^{-3}	3.73×10^{-3}	4.56×10^{-3}	6.18×10^{-3}	3.67×10^{-3}	3.95×10^{-3}
2600	3.70×10^{-3}	4.25×10^{-3}	5.21×10^{-3}	7.05×10^{-3}	4.18×10^{-3}	4.51×10^{-3}
2800	4.18×10^{-3}	4.80×10^{-3}	5.89×10^{-3}	7.96×10^{-3}	4.73×10^{-3}	5.09×10^{-3}
3000	4.69×10^{-3}	5.38×10^{-3}	6.61×10^{-3}	8.93×10^{-3}	5.29×10^{-3}	5.70×10^{-3}

expression proposed by Wilke (1950) which is strictly correct when only trace quantities are present

$$D_{H\text{-mix}} = \left[\sum_{i=1}^{n} \frac{x_j}{D_{H\text{-}j}} \right]^{-1}$$

The density of the mixture may be derived from the effective molecular weight of the combustion products on the basis of the ideal gas law.

Calculated values of the Lewis number for hydrogen atoms diffusing through stoichiometric hydrocarbon/oxygen combustion products decrease with increasing temperature, typical values being about 8 at 400 K and 4 at 3000 K. Since it has been assumed that equilibrium is maintained in the boundary layer, it is logical to use a mean value of the Lewis number between the flame temperature and the temperature when the hydrogen atom concentration becomes negligible (about 1700 K).

The following example illustrates the method for calculating the convective heat flux from an oxy-methane flame to a 10 mm diameter cylinder in cross flow.

Heat transfer from an oxy-methane flame

Data

Methane to oxygen ratio	=	1 : 2.04
Flame temperature	=	3000 K
Solid temperature	=	600 K
Cylinder diameter	=	0.01 m
Mass velocity of gases	=	3.524 kg $m^{-2}s^{-1}$ (obtained by dividing the mass flow rate of the methane + oxygen by the cross sectional area of the flame)

The compositions of the combustion products at temperatures between 600 K and 3000 K, calculated by the method given in Appendix 1 are given in in Table 3.12.

The specific heat, viscosity and thermal conductivity of the combustion products are calculated from the formulae given in Section 3.1, using the properties of the component gases in Tables 3.1-3.6. The diffusion coefficient of hydrogen atoms through the gas mixture is derived by the method given in Section 3.4.1. The resulting properties of the combustion products are given in Table 3.13. Using these values, the weighted means of the Prandtl and Reynolds number between 600 and 3000 K (Section 3.2) are:

$$\overline{Pr} = 0.68 \quad \text{and} \quad \overline{Re} = 666$$

The mean Lewis number (Section 3.4.1) between 1600 and 3000 K is:

$$Le = 4.7$$

The total enthalpy difference between 3000 and 600 K can be calculated from the compositions of the combustion products at these temperatures and the thermochemical data from the J.A.N.A.F. (1965) tables.

$$\Delta h = 8.782 \text{ MJ kg}^{-1}$$

The heat of recombination of hydrogen atoms (Section 3.4) is derived from the mole fraction of hydrogen atoms at 3000 K, the heat of recombination per mole (218.14 kJ mol^{-1}) and the mean molecular weight of the combustion products

$$\Delta h_{\text{chem}} = \frac{0.0398\ (\times\ 1)\ \times\ 218.14}{22.14}\ \text{J g}^{-1}$$

$$= 392.1 \text{ kJ kg}^{-1}$$

Table 3.12. Composition of methane/oxygen combustion products $CH_4 : O_2 = 1 : 2.04$

Component	Mole fractions						
	600 K	1000 K	1400 K	1800 K	2200 K	2600 K	3000 K
H_2O	0.6566	0.6566	0.6566	0.6554	0.6405	0.5727	0.4152
CO_2	0.3283	0.3283	0.3283	0.3275	0.3121	0.2428	0.1288
CO	0	0	0	0.0005	0.0126	0.0657	0.1430
O_2	0.0151	0.0151	0.0151	0.0153	0.0218	0.0498	0.0870
H_2	0	0	0	0.0003	0.0049	0.0244	0.0639
OH	0	0	0.00004	0.0009	0.0074	0.0335	0.0892
O	0	0	0	0	0.0004	0.0052	0.0332
H	0	0	0	0	0.0004	0.0060	0.0398

Table 3.13. Thermal and transport properties of combustion products

Temperature K	Mean Molecular Weight g	Density kg m^{-3}	Specific Heat kJ $kg^{-1}K^{-1}$	Viscosity $\mu Ns\ m^{-2}$	Thermal Conductivity mW $m^{-1}K^{-1}$	Diffusion Coefficient $m^2\ s^{-1}$
600	26.75	543.6	1.490	24.47	52.6	0.393×10^{-3}
1000	26.75	326.1	1.698	38.25	93.7	0.935×10^{-3}
1400	26.75	232.9	1.858	50.26	135.5	1.636×10^{-3}
1800	26.75	181.0	1.973	61.09	176.1	2.490×10^{-3}
2200	26.73	146.6	2.051	70.63	213.3	3.485×10^{-3}
2600	25.18	117.8	2.098	79.85	251.9	4.694×10^{-3}
3000	22.14	90.0	2.126	88.43	304.5	6.333×10^{-3}

For a cylinder in cross flow, McAdams (1954) gives the following convective heat transfer correlation:

$$Nu = 0.7\, Pr^{0.33}\, Re^{0.5}$$

This may be transformed in terms of Stanton number by the relation $St = Nu/Pr\, Re$ That is,

$$St = 0.7\, Pr^{-0.66}\, Re^{-0.5} = h/G\, c_{\mathrm{p}}$$

and the heat flux

$$q = 0.7\, Pr^{-0.66}\, Re^{-0.5}\, G\, c_{\mathrm{p}} \Delta T$$

Using the method given in Section 3.4, the corrected heat flux allowing for diffusion and recombiantion becomes:

$$q = 0.7\, \overline{Pr}^{-0.66}\, \overline{Re}^{-0.5} \left[1 + (Le-1)\frac{\Delta h_{\mathrm{chem}}}{\Delta h}\right]^{0.66} G\, \Delta h$$

$$= 1.2 \text{ MW m}^{-2}$$

This value can be compared with that calculated from the unmodified formula:

$$q = 0.7\, Pr^{-0.66}\, Re^{-0.5}\, G\, c_{\mathrm{p}}\, \Delta T$$

in which the properties are evaluated at the mean temperature between the flame gases and the solid surface (1800 K).

$Pr = 0.687$ $\quad Re = 577$

$C_{\mathrm{p}} = 1.973$ $\quad G = 3.524$

$\Delta T = 2400$

and $q = 0.62$ MW m^{-2}

It is apparent that the unmodified formula under-estimates the heat flux by almost 100%.

3.4.2 Experimental Measurements

Kilham and Purvis (1971) measured heat fluxes from premixed methane/oxygen and propane/oxygen flames of various equivalence ratios. The experimental results for heat transfer to a hemispherical nosed probe were compared with those calculated, assuming that: (a) recombination reactions of dissociated species are frozen in the boundary layer and the surface is noncatalytic, that is, a conventional low temperature correlation equation was used; (b) equilibrium is maintained throughout the boundary layer, but the Lewis number of all atoms and radicals is unity and (c) there is a Lewis number augmentation effect for hydrogen atoms as described above.

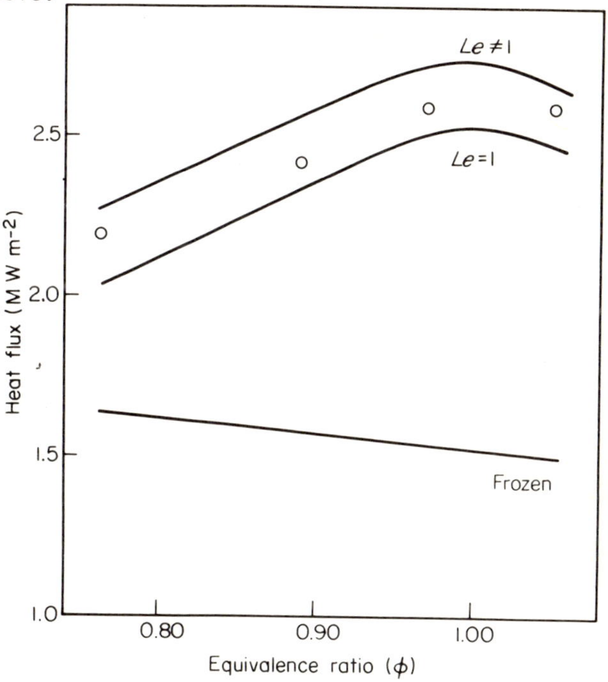

Figure 3.4   Heat fluxes for methane/oxygen flames of various equivalence ratios

These comparisons, which are illustrated in Figure 3.4 for methane/oxygen flames, showed that method (a) seriously underestimates the heat transfer and that the experimental values are intermediate between those calculated on the basis of assumptions (b) and (c).

When computing convective heat transfer from high temperature fuel/oxygen flames it would appear that the use of equations of the type of Equation 3.25 is, therefore, justified.

If the surface is very close to the actual flame reaction zone as in the earlier work of Kilham *et al.* (1963, 1967), there is some evidence to indicate that the concentrations of hydrogen atoms and free radicals are higher than those calculated on an equilibrium basis. This can lead to a further enhancement of the heat transfer rate, the extent of which is difficult to assess because of the lack of knowledge of high temperature reaction kinetics.

4 Radiative Heat Transfer

When a body has a temperature above absolute zero, it emits energy continuously at a rate which is a function of its temperature. This energy is called thermal radiation and it is transmitted by electromagnetic waves within the wavelength range 0.1 to 100 micrometre (μm). These wavelengths only constitute a small part of the spectrum over which energy is transmitted by electromagnetic waves. An indication of the extent of the electromagnetic spectrum is shown in Figure 4.1.

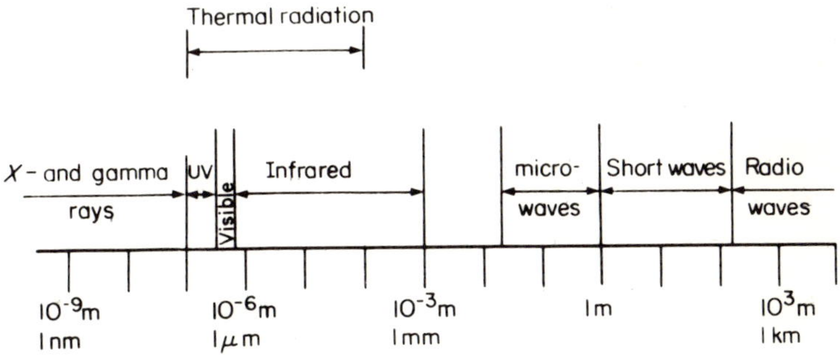

Figure 4.1 Electromagnetic spectrum

In simple terms, thermal radiation can be said to travel in straight lines with the velocity of light, so that opaque objects cast shadows when placed in the path of thermal radiation and one body cannot receive radiation directly from another unless it can 'see it'. When a body receives thermal radiation from another, it absorbs some of it, reflects some and transmits the remainder.

$$\alpha + \rho + \tau = 1 \tag{4.1}$$

where α, ρ and τ* are the fractions of received radiation which are absorbed, reflected and transmitted respectively. Reflection of radiation from the surface of the body may occur in such a way that the angle of reflection equals the angle of incidence. In this case, reflection is called specular or regular. Reflection from rough surfaces occurs indiscriminately in all directions, that is, is diffusely radiated.

4.1 MAXIMUM RADIATION

A black body is defined as one which emits the maximum possible amount of radiation corresponding to its temperature; other bodies emit less. The rate at which radiation is emitted per unit area (radiative flux) by a black body over the whole thermal range of wavelengths is given by the Stefan-Boltzmann law:

$$E_b = \sigma T^4 \tag{4.2}$$

where σ is the Stefan-Boltzmann constant equal to 56.7×10^{-12} kW m^{-2} K^{-4} and T is the absolute temperature of the body.

Calculations can be conveniently carried out by using Equation 4.2 in the form:

$$E_b = 56.7 \times (T/1000)^4 \text{ kW m}^{-2} \tag{4.3}$$

For example, the energy radiated from a black surface at a temperature of 1000 K (red hot) is 56.7 kW m^{-2}.

*Absorptivity, reflectivity and transmissivity respectively.

4.2 VARIATION OF THERMAL RADIATION WITH WAVELENGTH

The rate at which radiation is emitted by a black body varies with wavelength λ as shown in Figure 4.2; that emitted at a particular wavelength (monochromatic radiation) is given by Planck's distribution law:

$$E_{b\lambda} = \frac{c_1 \lambda^{-5}}{\exp(c_2/\lambda T) - 1} \qquad (4.4)$$

where $E_{b\lambda}$ is the monochromatic emissive power of a black body and the Planck's constants c_1 and c_2 equal 3.74×10^{-19} kW m^2 and 1.439×10^{-2} m K, respectively.

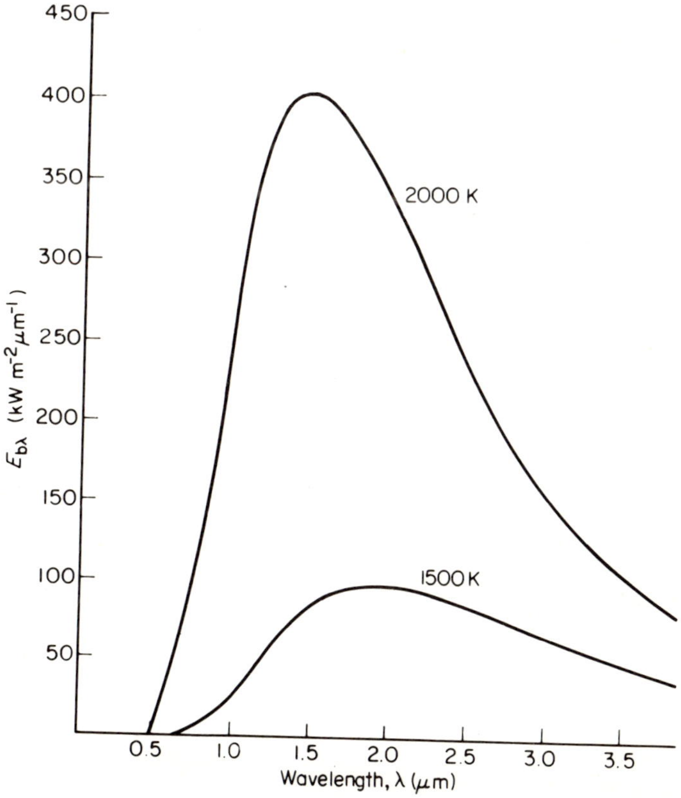

Figure 4.2 Radiation emitted by a black body

Using the approximation that exp(x) - 1 = exp(x) for large values of x, Wein's law may be obtained from Equation 4.4:

$$E_{b\lambda} = c_1/\left[\lambda^5 \exp(c_2/\lambda T)\right] \qquad (4.5)$$

Wein's law is in error by about 1% when λT rises to 0.003 m K.

Equation 4.4 may be rewritten as:

$$E_{b\lambda}/T^5 = \frac{c_1(\lambda T)^{-5}}{\exp(c_2/\lambda T) - 1} \qquad (4.6)$$

to show that a plot of $E_{b\lambda}/T^5$ against λT produces a single curve, whereas the usual plot of $E_{b\lambda}$ against λ produces a family of curves, one for each temperature.

Figure 4.2 shows that a black body emits more energy as its temperature increases and also that more of that energy is emitted at shorter wavelengths. The wavelength at which the maximum amount of radiation is emitted from a black body is given by Wien's displacement law:

$$\lambda_{max}.T = 2.9 \times 10^{-3} \text{ m K} = 2900 \text{ µm K} \qquad (4.7)$$

Thus, the wavelengths at which maximum radiation is emitted from the sun and earth, which may be considered to be black bodies at temperatures of 5800 and 288 K, are in the middle of the visible range at 0.5 µm and well into the infra-red at 10 µm, respectively.

4.3 INTENSITY OF RADIATION

The Stefan-Boltzmann law (Equation 4.2) gives the amount of energy emitted from a black surface into all the space above it (kW m^{-2}). The intensity of radiation (I) is defined as that emitted per unit solid angle (kW m^{-2} $ster^{-1}$). Lambert's law states that the intensity of radiation emitted at an angle ϕ to the normal to the surface may be related to that emitted normal to the surface by:

$$I_\phi = I_n \cos\phi \qquad (4.8)$$

The relationship between the intensity of radiation normal to the surface (I_n) and the black body emissive power ($E_b = \sigma T^4$) can be derived from considering the radiation from the elemental area, dA, in Figure 4.3. A strip on the hemispherical envelope may be defined which is at an angle ϕ to the normal and which subtends an angle $d\phi$ at dA. The solid angle subtended by the strip at dA is the area of the strip divided by the square of the radius, that is, $2\pi r \sin\phi (r\, d\phi)/r^2 = 2\pi \sin\phi\, d\phi$ and the radiation directly incident on the strip from dA is $I_n \cos\phi\, 2\pi \sin\phi\, d\phi$. Integrating this expression from 0 to $\pi/2$ gives the total radiation emitted from dA into the hemisphere of space above it. The Stefan-Boltzmann law also gives this quantity so that:

$$\sigma T^4\, dA = 2\pi\, I_n\, dA \int_0^{\pi/2} \sin\phi \cos\phi\, d\phi$$

$$= 2\pi\, I_n\, dA \int_0^{\pi/2} \tfrac{1}{2} \sin 2\phi\, d\phi$$

$$= \pi\, I_n\, dA$$

and

$$I_n = \sigma T^4/\pi \qquad (4.9)$$

4.4 INVERSE SQUARE LAW

A small cone (angle, $\Delta\omega$) of radiation emitted from a surface of area ΔA intercepts an area $\ell^2\Delta\omega$ on a sphere of radius ℓ which has its centre at the source of radiation (Figure 4.4). Since this area is proportional to ℓ^2, then the radiation received per unit area is inversely proportional to ℓ^2, a relationship known as the inverse square law.

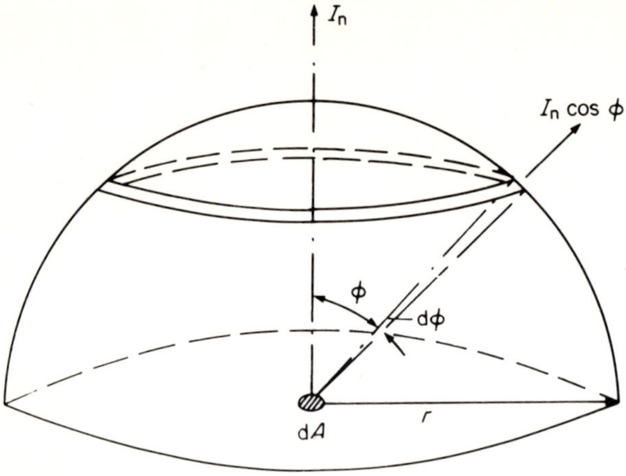

Figure 4.3 Intensity of radiation

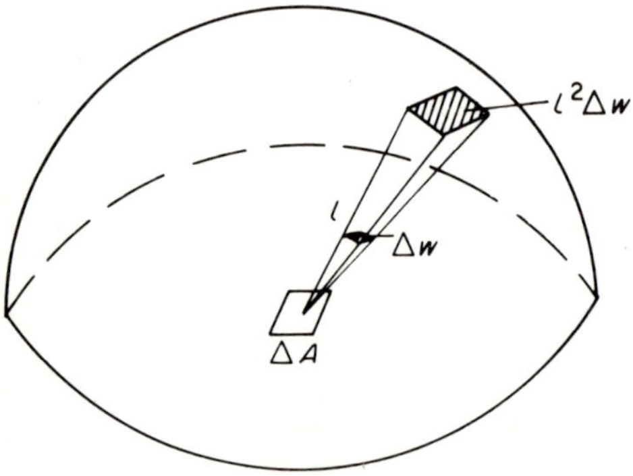

Figure 4.4 Radiation emitted through a solid angle

4.5 EMISSIVITY

The emissivity (ε) of a body or surface is the ratio of the radiative flux (E) emitted by the body to that emitted (E_b) by a black body at the same temperature. By definition, ε is less than 1 for any body which is not black. Emissivities can be defined by the whole range of thermal wavelengths or at specific wavelengths:

$$\varepsilon = E/E_b \quad \text{or} \quad \varepsilon_\lambda = E_\lambda/E_{b\lambda} \qquad (4.10)$$

A grey body is defined as one whose emissivity is constant with wavelength; the energy radiated from it per unit area is:

$$E = \varepsilon\sigma T^4 = \varepsilon E_b \qquad (4.11)$$

Like a black body, a grey body is an ideal which can be approached by some real bodies. In general, the emissivity of a real body or surface can vary with wavelength, angle of emission and temperature and a more precise definition of emissivity is:

$$\varepsilon_{\lambda\phi T} = E_{\lambda\phi T}/E_{b\lambda T} \qquad (4.12)$$

Similarly, the absorptivity, reflectivity and transmissivity of a surface can vary with wavelength, angle of incidence of radiation and also the temperature of the radiation, in which case they may be represented by $\alpha_{\lambda\phi T}$, $\rho_{\lambda\phi T}$ and $\tau_{\lambda\phi T}$. Figure 4.5 shows the emissivity of aluminium as a function of wavelength.

4.6 KIRCHOFFS LAW

Kirchoffs law relates the absorptivity and emissivity of a body or surface; it states that the monochromatic emissivity of a surface is equal to its monochromatic absorptivity:

$$\varepsilon_\lambda = \alpha_\lambda \qquad (4.13)$$

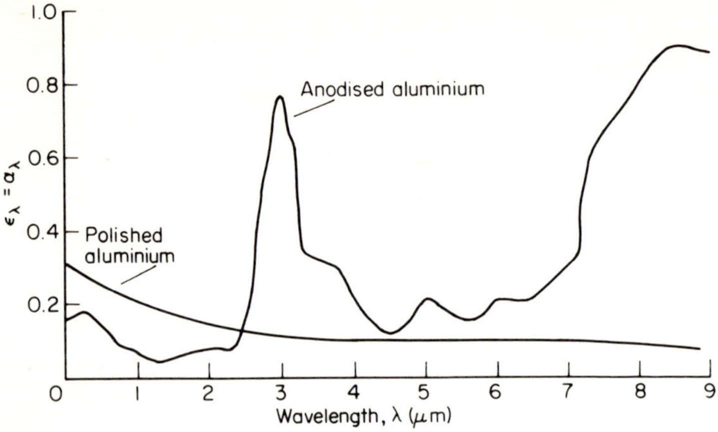

Figure 4.5 Emissivity of aluminium

Since the emissivity of a grey body is constant with wavelength, the overall emissivity is equal to overall absorptivity:

$$\varepsilon = \alpha \tag{4.14}$$

For a body whose emissivity (and absorptivity) varies with wavelength, the average emissivity is obtained by integrating the emission at each wavelength over all wavelengths and dividing by the black body emission. From Equation 4.10,

$$\varepsilon = \frac{E}{E_b} = \frac{\int_0^\infty \varepsilon_\lambda E_{b\lambda} \, d\lambda}{E_b}$$

$$= \int_0^\infty \varepsilon_\lambda E_{b\lambda} \, d\lambda \, / \, \sigma T^4 \tag{4.15}$$

where T is the temperature of the emitting surface.

Similarly,

$$\alpha = \int_0^\infty \alpha_\lambda \; E'_{b\lambda} \; d\lambda \, / \, \sigma T'^4 \qquad (4.16)$$

where T' is the temperature of the black body source. Unless $T = T'$, the overall emissivity does not equal the overall absorptivity:

$$\varepsilon \neq \alpha \qquad (4.17)$$

4.7 TEMPERATURES

The temperature of a black body can be determined by measuring the amount of radiation it emits (E_b) and calculating the temperature from:

$$T = (E_b/\sigma)^{\frac{1}{4}} \qquad (4.18)$$

If the body is not black and emits q kW m^{-2}, an effective black body (or brightness) temperature of the body can be defined by

$$T_b = (q/\sigma)^{\frac{1}{4}}$$

In this case, $q = \sigma T_b^4$ and it becomes unnecessary to evaluate the emissivity.

4.8 ABSORPTION OF RADIATION BY AN ABSORBING MEDIUM

When radiation of intensity I passes through a very thin layer of an absorbing medium (Figure 4.6), it is attenuated by an amount (dI) which is proportional to the intensity of radiation and to the path length ($d\ell$):

$$- dI = K \, I \, d\ell \qquad (4.19)$$

where K is the attenuation or extinction coefficient. Attenuation of radiation of intensity I_0 incident upon a thick layer of an absorbing medium (Figure 4.6) is given by the integration of Equation 4.19

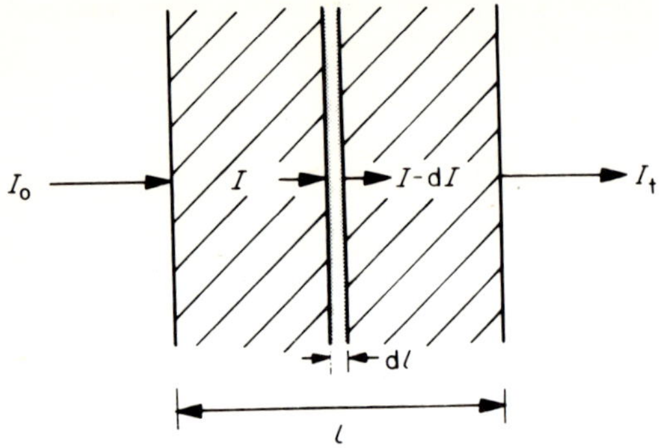

Figure 4.6 Attenuation of radiation by an absorbing medium

between the limits I_0 and I_t, and 0 and ℓ:

$$I_t = I_0 \exp\left(\int_0^{\ell} - K \, d\ell\right) \tag{4.20}$$

If K can be assumed to be independent of ℓ:

$$I_t = I_0 \exp(-K\ell) \tag{4.21}$$

and the transmissivity is:

$$\tau = \exp(-K\ell) \tag{4.22}$$

K has a value of the order of 1 m^{-1} for a pulverised fuel flame.

The fraction of the radiation which is not transmitted is $1 - \tau = \alpha$, the absorptivity:

$$\alpha = 1 - \exp(-K\ell) \tag{4.23}$$

For a grey medium,

$$\varepsilon = \alpha = 1 - \exp(-K\ell) \tag{4.24}$$

If the absorbing medium is a mixture of gases, it is usual to express K for each gas as a function of the partial pressure of that gas:

$$K = k\, p \qquad (4.25)$$

where p is the partial pressure of the gas,

and $$I_t/I_0 = \exp(-k\, p\, \ell) \qquad (4.26)$$

Mean beam lengths

When radiation is transmitted from one surface to another through an intervening medium, the paths being traversed by radiation from different positions on the surfaces are of different lengths. A mean length can be chosen to describe the overall transmission from one surface to the other, so that the transmissivity is:

$$\tau_{12} = \exp(-K\, L_{12}) \qquad (4.27)$$

where L_{12} is the mean beam length between surface 1 and surface 2. Similarly, emissivity and absorptivity of an absorbing medium with respect to a surface can be expressed by:

$$\varepsilon_m = 1 - \exp(-K\, L_{ms}) \qquad (4.28)$$

where L_{ms} is a mean beam length between medium m and surface s. Consequently, the radiation emitted by the medium to surface s is:

$$q = \left[1 - \exp(-K\, L_{ms})\right] \sigma T^4 \qquad (4.29)$$

$$= \varepsilon_m\, \sigma\, T^4 \qquad (4.30)$$

$$= \varepsilon_m\, E_{bm} \qquad (4.31)$$

Table 4.1 is a list of values of L_{ms} which can be used to determine the emission of radiation from the absorbing medium to a surface. A single value

Table 4.1. Mean beam lengths (absorbing medium to a surface)

Shape	Characterising dimension	Mean beam length / Characteristic length
Right circular cylinders	*Diameter*	
(1) Height = diameter, radiating to whole surface		0.6
(2) Height = 0.5 × diameter, radiating to		
(a) end		0.43
(b) side		0.46
(c) whole surface		0.45
(3) Height = 2 × diameter, radiating to		
(a) end		0.60
(b) side		0.76
(c) whole surface		0.73
Rectangular parallelopipeds	*Shortest edge*	
(1) Cube, radiating to side		0.60
(2) 1 × 1 × 4 radiating to		
(a) 1 × 4 face		0.82
(b) 1 × 1 face		0.71
(c) all faces		0.81

may be used if radiation from the absorbing medium to the entire bounding surface is required.
It is:

$$L_{ms} = 3.5\ V/A \qquad (4.32)$$

where V is the volume of the medium and A is its surface area.

4.9 INTERCHANGE OF RADIATION BETWEEN TWO SURFACES

The amount of radiation emitted by one surface which is incident on another depends on the view which the surfaces have of each other. It is usual to express this quantity in terms of a view factor (F), which is defined as the fraction of the total radiation emitted from one surface which is directly incident on the other (Figure 4.7). Thus, the amount of radiation leaving a grey surface 1 is $\varepsilon_1 A_1 E_{b1}$ and that directly incident upon surface 2 is $(\varepsilon_1 A_1 E_{b1}) \times F_{12}$.

For an enclosure of n surfaces:

$$F_{11} + F_{12} + F_{13} + \ldots F_{1n} = 1 \qquad (4.33)$$

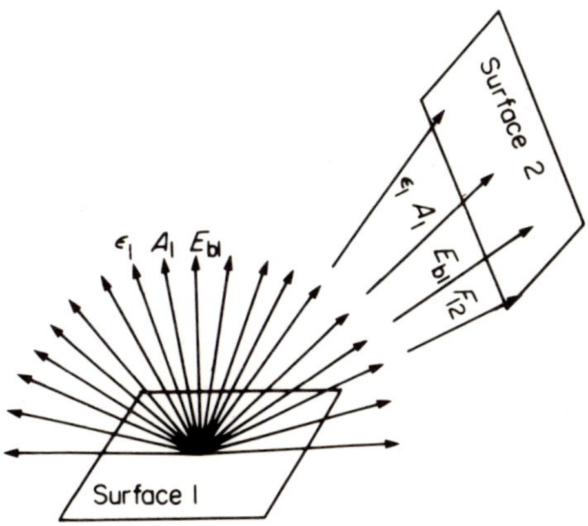

Figure 4.7 Radiation from one surface directly incident upon a second

4.10 EVALUATION OF VIEW FACTORS

The view factor, F_{12}, may be expressed mathematically by considering the elemental areas dA_1 and dA_2 which are part of the two black surfaces A_1 and A_2 (Figure 4.8). dA_2 subtends the solid angle $d\omega_1 = dA_2 . \cos \phi_2/\ell^2$ at dA_1. The intensity of radiation from dA_1 directed towards dA_2 is $dA_1\ E_1 \cos \phi_1/\pi$. Consequently, the energy emitted from surface A_1 which is directly incident upon surface A_2 is:

$$Q = E_1 \int_{A_1}\int_{A_2} \cos \phi_1\ dA_1 \cos \phi_2\ dA_2/\pi\ell^2 \quad (4.34)$$

$$= E_1\ A_1\ F_{12}$$

$$\text{or} \quad A_1\ F_{12} = \int_{A_1}\int_{A_2} \cos \phi_1\ dA_1 \cos \phi_2\ dA_2/\pi\ell^2 \quad (4.35)$$

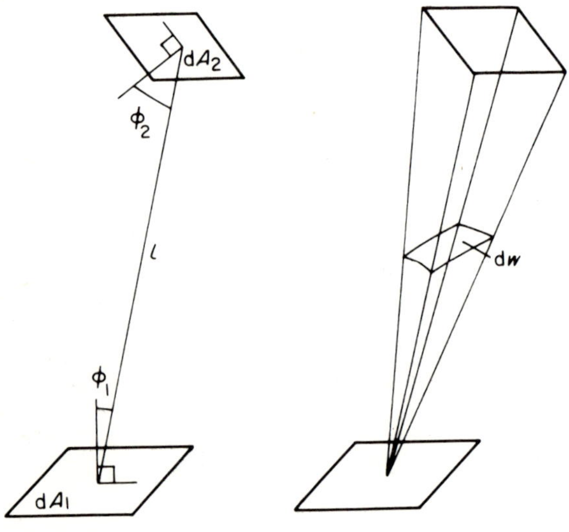

Figure 4.8 Direct interchange of radiation between two surfaces of differential area

Similar consideration of the energy emitted from surface A_2 or examination of the symmetry of Equation 4.35 shows:

$$A_1 F_{12} = A_2 F_{21} \quad \text{or} \quad s_1 s_2 = s_2 s_1 \quad \text{or} \quad \overline{12} = \overline{21} \tag{4.36}$$

where $A_1 F_{12}$ (or $s_1 s_2$ or $\overline{12}$) is the exchange area between surfaces 1 and 2.

Evaluation of Equation 4.35 for a particular situation usually involves double integration over length instead of areas:

$$A_1 F_{12} = \int_x \int_y \int_{x'} \int_{y'} \cos \phi_1 \, dx \, dy \cos \phi_2 \, dx' dy' / \pi \ell^2 \tag{4.37}$$

where $dA_1 = dx\, dy$ and $dA_2 = dx'\, dy'$. Solution of this integral equation is not always simple and consequently data on values of the view factors for a number of geometries have been tabulated or plotted [Kreith (1962), Hottel and Sarofim (1967), Leuenberger and Person (1956), Sparrow and Cess (1966), Wiebelt (1966)]. Figures 4.9 and 4.10 show charts based on the geometries indicated.

For geometries where neither tables nor charts of view factors exist, it is necessary to evaluate Equation 4.35. Equation 4.35 is expressed as a double integral but since (in cartesian co-ordinates) dA may be expressed as the product $dx\, dy$, then, strictly, the integral is quadruple. In general, analytical solutions are not possible and numerical methods have to be used. These generally extend the concept of an integral of a function of a single variable being the area under a curve to four'dimensions'. A possible method is one of the Newton-Stokes quadrature formulae such as the trapezoidal rule or Simpson's rule. Alternatively, Monte Carlo integration could be carried out.

4.11 INTERCHANGE OF RADIATION AMONGST MORE THAN TWO SURFACES

If more than two surfaces are being examined, both indirect and direct interchange amongst them

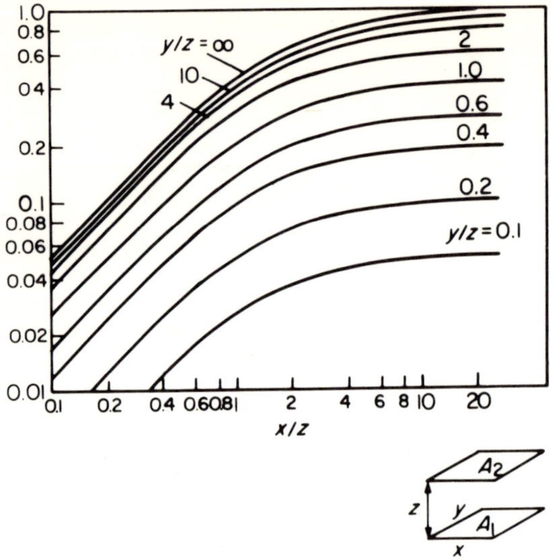

Figure 4.9 View factor chart

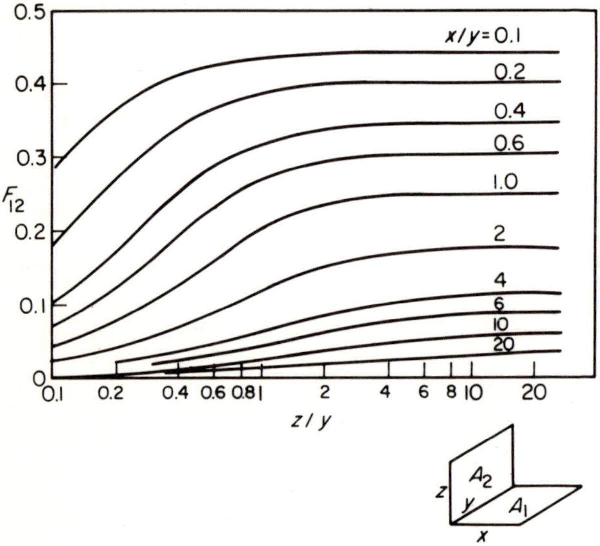

Figure 4.10 View factor chart

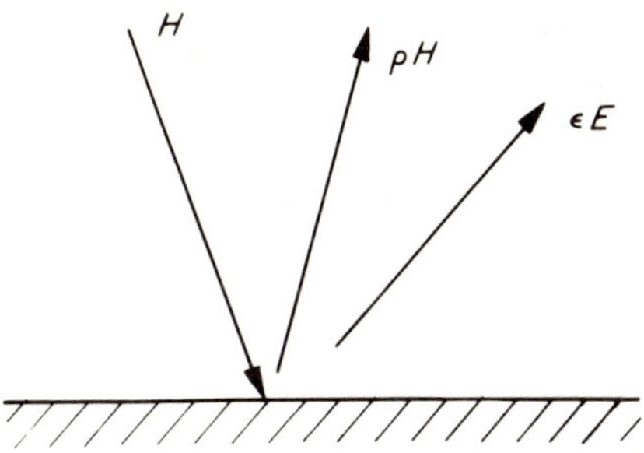

Figure 4.11 Radiation leaving a grey surface

must be considered. It is convenient to examine the simplest case of interchange amongst grey surfaces, each of a constant temperature. Radiation from each surface is also assumed to be diffuse. The analysis of heat transfer in this situation can be made by considering the radiation leaving each surface.

The radiation incident upon (H) and leaving (W) unit area of a grey surface is shown in Figure 4.11. That leaving is the sum of the reflected radiation (ρH) and that which is emitted by the surface by virtue of its temperature (εE_b).

$$W = \varepsilon E_b + \rho H \qquad (4.38)$$

W, the leaving flux, is often called the radiosity. The net transfer of radiation from the surface is:

$$q = W - H$$

$$= W - \frac{(W - \varepsilon E_b)}{\rho} = \frac{\varepsilon E_b - W(1 - \rho)}{\rho}$$

For a grey body, $\varepsilon = \alpha = 1 - \rho$ and,

$$q = \varepsilon(E_b - W)/\rho \qquad (4.39)$$

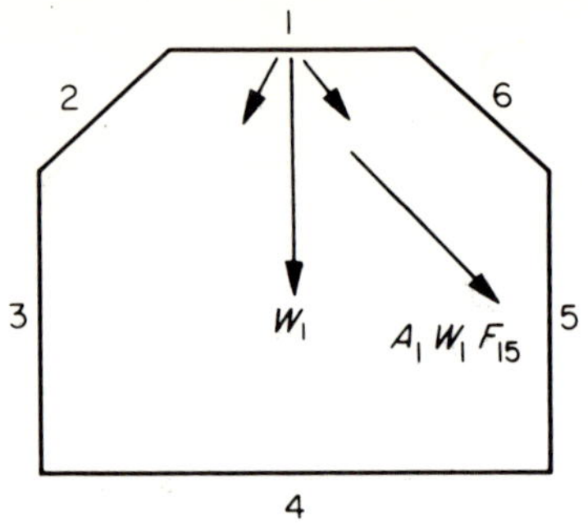

Figure 4.12 Enclosure comprising grey surfaces

Equation 4.38 may be applied to each of the *n* surfaces comprising an enclosure (Figure 4.12).

For any surface i, the radiation leaving unit area is:

$$W_i = \varepsilon_i E_{bi} + \rho_i H_i \tag{4.40}$$

The total energy incident upon this surface is the sum of the energy leaving every other surface multiplied by a view factor.

$$A_i H_i = W_1 A_1 F_{1i} + W_2 A_2 F_{2i} + \dots W_n A_n F_{ni} \tag{4.41}$$

This procedure completely takes into account reflections within the enclosure and it is not necessary to use modified view factors like $\mathfrak{F}$ adopted by some investigators (Hottel in McAdams, 1954). Because

$$A_i F_{ij} = A_j F_{ji}$$

$$H_i = W_1 F_{i1} + W_2 F_{i2} + \dots W_i F_{ii} + \dots W_n F_{in}$$

Or,

$$H_i = \sum_{j=1}^{n} W_j F_{ij} \tag{4.42}$$

Substituting this expression for the incident radiation into Equation 4.40 gives:

$$W_i = \varepsilon_i E_{bi} + \rho_i \left(\sum_{j=1}^{n} W_j F_{ij} \right) \qquad (4.43)$$

where i = 1, 2, 3 ... n. These n linear simultaneous equations (Equation 4.43) can be solved to provide values of leaving radiation for every surface. The net heat transferred from each surface is:

$$Q_i = \frac{\varepsilon_i \; (E_{bi} - W_i) \; A_i}{(1 - \varepsilon_i)} \qquad (4.44)$$

The surfaces discussed so far may comprise an enclosure such as a furnace. In this case, the furnace enclosure usually contains an absorbing medium such as a flame or combustion products as well as the stock which is being heated.

4.12 HEAT TRANSFER IN AN ENCLOSURE CONTAINING A FLAME

Equations of the kind given by Equation 4.43 are still appropriate for the radiation leaving each surface of an enclosure containing a flame or combustion products. It is merely necessary to take account of two further factors. They are the attenuation of radiation by the flame as it is transmitted from one surface to another and the radiation emitted by the flame. The energy incident upon each surface is:

$$H_i = W_1 F_{i1} \tau_{i1} + W_2 F_{i2} \tau_{i2} + \ldots W_n F_{in} \tau_{in} + \varepsilon_f \sigma T_f^4 \qquad (4.45)$$

$$= \sum_{j=1}^{n} W_j F_{ij} \tau_{ij} + \varepsilon_f \sigma T_f^4$$

where τ_{ij} is the transmissivity of the flame between surfaces i and j and the flame is assumed to be grey at a temperature T_f. Substituting the value of H_i from Equation 4.45 into Equation 4.40 gives n linear simultaneous equations as before, with the radiation leaving each surface as the unknown:

$$W_i = \varepsilon_i E_{bi} + \rho_i \sum_{j=1}^{n} W_j F_{ij} \tau_{ij} + \rho_i \varepsilon \sigma T_f^4 \qquad (4.46)$$

The transmissivities and the emissivity of the flame can be expressed in terms of mean beam lengths but it is important to note that different values of mean beam lengths must be used for different surface-to-surface configurations and for the emissivity.

5 Radiative Heat Transfer from Flames

The flame itself and the combustion products absorb, and emit, thermal radiation. Both gases and particulate material present in the flame contribute to the absorbing propensity of the flame.

5.1 GASES

Many gaseous species are present in flames from hydrocarbon fuels including CO_2, H_2O, CO, H_2, O_2, H, O, OH, N_2, N and NO which have already been referred to. The only non-absorbing constitutents are the atomic species (H, O, N) and the homonuclear diatomic molecules (H_2, O_2) whilst the rest absorb, and emit, radiation over specific wavelength ranges or bands within the thermal spectrum. Figure 5.1 shows the absorption characteristics of carbon dioxide. Carbon dioxide and water vapour are the most important absorbing gases in combustion products. Table 5.1 shows the ranges of wavelength over which absorption takes place. Often, carbon dioxide and water vapour are the only two gases taken into account in engineering calculations, but more precise calculations demand the inclusion of other gases (Hottel and Sarofim, 1967).

Although the spectral absorptivities of the absorbing gases in combustion products vary enormously with wavelength, an average value of absorptivity or emissivity for the whole of the wavelength range under consideration is normally used. this procedure is equivalent

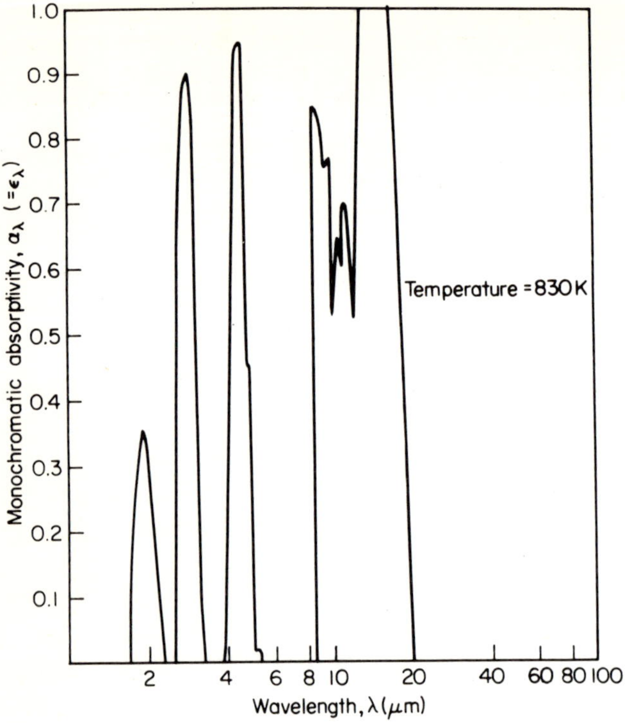

Figure 5.1 Absorption of thermal radiation by carbon dioxide

Table 5.1. Ranges of wavelengths over which carbon dioxide and water vapour absorb thermal radiation

	Range of wavelengths, μm		
Carbon dioxide	2.4 to 3.0	4.0 to 4.8	12.5 to 16.5
Water vapour	2.2 to 3.3	4.8 to 8.5	12 to 25

to making the assumption that the gas is grey. The average values of emissivity and absorptivity have been obtained by a number of experimentalists and expressed in terms of the following variables:

$$\varepsilon_{gs} = \varepsilon(L_{gs}, p, P, C, T_g) \tag{5.1}$$

and

$$\alpha_{gs} = \alpha_{gs}(L_{gs}, p, P, C, T_g, T_s) \tag{5.2}$$

where L_{gs} is the mean beam length for the gas to the surface, p the partial pressure of the constituent being considered, P the total pressure of the combustion gases, C the composition of other gases in the combustion gases, T_g the temperature of the combustion gases and T_s the temperature of the surface from which the radiation is received by the combustion gases.

The emissivity of the combustion products arising from carbon dioxide and water vapour may be calculated from:

$$\varepsilon_g = \varepsilon_{CO_2} + \varepsilon_{H_2O} - \Delta\varepsilon$$

$$= \varepsilon^*_{CO_2} C_{CO_2} + \varepsilon^*_{H_2O} C_{H_2O} - \Delta\varepsilon \tag{5.3}$$

where $\varepsilon^*_{CO_2}$ is emissivity of carbon dioxide at a total pressure of 1 bar, C_{CO_2} is the correction factor for total pressures different from 1 bar, $\varepsilon^*_{H_2O}$ is the emissivity of water for a total pressure of 1 bar and in the limit as the partial pressure of water approaches zero, C_{H_2O} is the correction for total pressure different from 1 bar and partial pressure of water different from zero, and $\Delta\varepsilon$ is the correction factor for the overlapping of the carbon dioxide and water bands.

Similarly, the absorptivity of the combustion products arising from carbon dioxide and water vapour may be calculated from:

$$\alpha_g = \alpha_{CO_2} + \alpha_{H_2O} - \Delta\alpha \tag{5.4}$$

But the absorptivities may also be expressed in terms of the wall and gas temperatures (T_w and T_g), as follows:

$$\alpha_{CO_2} = \varepsilon^*_{CO_2} (T_g/T_w)^{0.65} C_{CO_2} \qquad (5.5)$$

and

$$\alpha_{H_2O} = \varepsilon^*_{H_2O} (T_g/T_w)^{0.45} C_{H_2O} \qquad (5.6)$$

$\Delta\alpha$ is evaluated at the temperature of the wall.

Data for calculations based on Equation 5.3 have been provided in graphical form by a number of workers (Hottel and Sarofim, 1967). The emissivity of the combustion products from methane/air and petroleum fuel oils/air can be obtained more simply from the data of Hadvig (1970) presented in Figures 5.2 and 5.3. Hadvig used values of the ratios of the partial pressures of carbon dioxide to water vapour of 2 : 1 and 1 : 1 for methane/air and petroleum fuel oil/air respectively. In this way, the emissivity of the combustion products, which stems from the amount of carbon dioxide and water vapour present, can be obtained merely by using the partial pressure of carbon dioxide.

For example, for a partial pressure of carbon dioxide of 0.075 bar in a petroleum fuel flame with a mean beam length of 4 m and a temperature of 1800 K:

$$p_{CO_2} L = 0.075 \times 4 = 0.3$$

Reading from Figure 5.3 with values of T = 1800 K and pL = 0.3, the emissivity of the flame is 0.19.

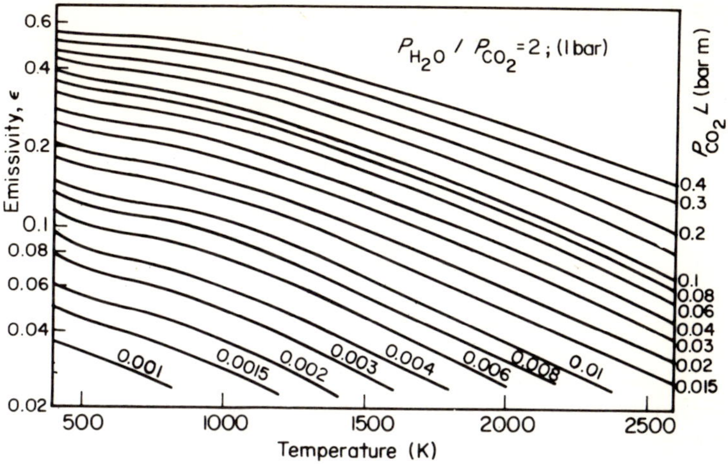

Figure 5.2 Emissivities of atmospheres containing carbon dioxide and water in the ratio 2 : 1

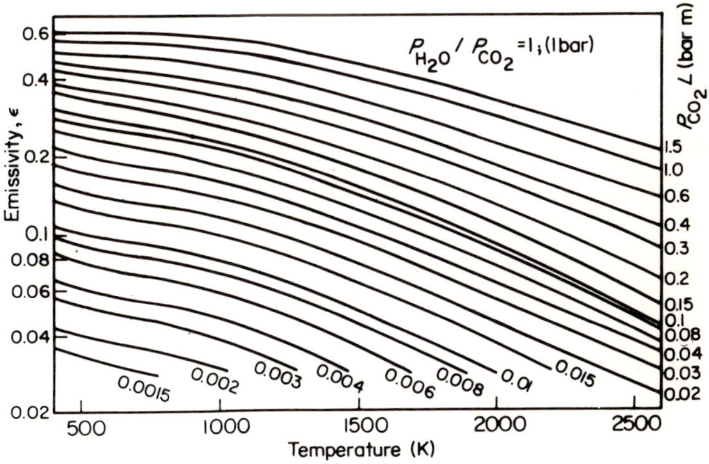

Figure 5.3 Emissivities of atmospheres containing carbon dioxide and water vapour in the ratio 1 : 1

5.2 PARTICLES IN A FLAME

When a beam of radiation is incident upon a suspension of particles, some of it is transmitted, some absorbed and some scattered. The scattered radiation includes that diffracted, refracted and reflected by the particles. The absorbed radiation is retransmitted at a wavelength corresponding to the temperature of the particle and not the source.

These interactions of radiation with particles depend critically on the particle size which is usually expressed as a ratio to the wavelength of the radiation being considered.

The attenuation of radiation after it has traversed a suspension of particles of diameter larger than about twice the wavelength of the radiation can be described by relatively simple theory based on absorption and diffraction. For very small particles of diameter less than a tenth of the wavelength, simple scattering theory can be used but, for particles of an intermediate size, more complex calculations are required to relate attenuation with the properties of the particles and their concentration.

A flame may contain particles of burning fuel, soot, char, or ash which, together, cover the following three categories of size with regard to thermal radiation.

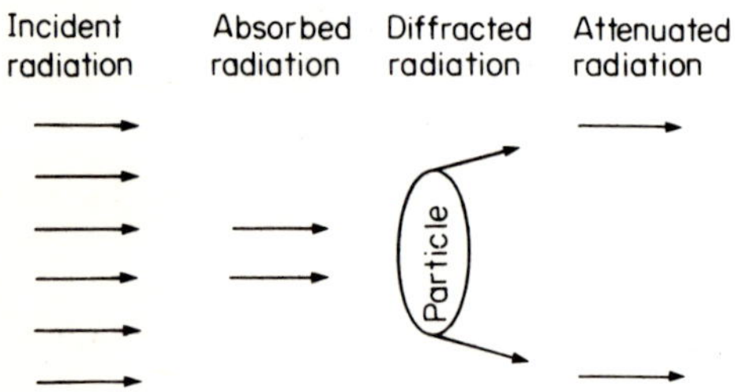

Figure 5.4 Attenuation of radiation by a large particle

5.2.1 Large Particles

Figure 5.4 illustrates what happens when radiation is incident upon a black particle suspended in a transmitting medium:

1. Radiation directly incident upon the particle is absorbed by it. For a spherical particle, the absorbing cross-sectional area is $\pi d^2/4$.
2. Radiation incident upon the boundary between the particles and the surrounding medium is diffracted by an amount proportional to the cross-sectional area or $\pi d^2/4$ for a spherical particle.
3. Radiation not absorbed or diffracted is transmitted. The incident radiation is thus attenuated by an amount proportional to $\pi d^2/4$.

But the diffracted radiation lies in the forward direction at an angle to the incident beam which depends on the particle size. So, depending on the particle size and angle of collection, the effective cross-sectional area for attenuation can vary from $\pi d^2/4$ to $2\pi d^2/4$. If F_{at} is defined as the ratio of the effective or apparent cross-sectional area to its geometrical cross-sectional area, F_{at} varies from 1 to 2.

Rose (1952) showed that, for an angle of collection of 1°, $F_{at} = 1$ for particles larger than 80 μm. When the angle of collection is about 10°, $F_{at} = 1$ for particles larger than about 25 μm.

The attenuation of radiation by a suspension of black particles ($F_{at} = 1$) of low concentration is simply the product of the number of particles per unit volume of gas (c_n) and the projected area per particle in the direction of the radiation (A_n):

$$A = c_n A_n \qquad (5.7)$$

Alternatively,

$$A = c_m A_m \qquad (5.8)$$

$$A = c_v A_v \qquad (5.9)$$

where subscripts m and v refer to concentration and area measured on a mass and volume basis respectively.

Thus,

$$K = c_v A_v \quad \text{and}$$

$$\varepsilon = 1 - \exp(-c_v A_v L) \qquad (5.10)$$

where $c_v A_v = K$ the absorption coefficient.

The absorption coefficient is often expressed in terms of the surface area per unit volume (S_v) rather than the projected area. Since $A_v = S_v/4$ for randomly disposed particles, the absorption coefficient of the suspension is:

$$K = c_v S_v/4 \qquad (5.11)$$

Further, if the particles are spherical, the surface area per unit volume is related to the diameter of the particles by $S_v = 6/d$, and

$$K = 3c_v/2d \qquad (5.12)$$

Thus, the emissivity of a suspension of particles of 100 µm in diameter and at a concentration of 2×10^{-5} m^3 m^{-3} contained in an enclosure of mean beam length of 1 m is

$$\varepsilon = 1 - \exp\left[-3 \times 2 \times 10^{-5}/(2 \times 10^{-4})\right]$$

$$= 1 - \exp(-0.3)$$

$$= 1 - 0.741$$

$$= 0.239$$

It is usual to assume that burning coal and char particles derived from coal or heavy fuel oil droplets in a flame are large convex black bodies for which $F_{at} = 1$ and $K = c_v S_v/4 = c_m S_m/4$.

Field *et al.* (1967) have shown, on the basis of the few data on optical properties that are available, that the assumption that burning coal or char are black is unlikely to give absorption cross-sections which are in error by more than 20% and, in general, the error will be much less. They have pointed out that pulverised fuel commonly has a specific surface of about 250 m^2 kg^{-1}.

When dispersed in air in stoichiometric proportions, its concentration at 1800 K is about 0.016 kg m^{-3} which gives a value of K 1.0 m^{-1}. In the primary jet, the absorption coefficient will be higher owing to the change in mass proportions of air and fuel and to the lower temperature. It could be as large as 30 m^{-1} at the burner.

During thermal decomposition of a non-swelling coal, before combustion begins S_m increases because of the reduction in particle density, but there is a compensating decrease in c_m and K remains approximately constant.

5.2.2 Small Particles (diameter less than approximately λ/10)

A simplified treatment of the influence of electromagnetic waves on small particles is possible if the particle is assumed to consist of a dielectric material (so that the field can penetrate it without attenuation) and also that the particle is sufficiently small for the field to be essentially uniform through it. A simple analysis based on the electric field experienced by the electrons in the particle provides the Rayleigh equation (Equation 5.13) for the intensity of scattered radiation.

The radiant flux (q_0 kW m^{-2}) incident upon a dielectric or non-absorbing sphere is scattered in the direction θ by an amount q, which is the sum of two incoherent, plane polarised components, i_1 and i_2:

$$q = A(1 + \cos^2\theta) = i_1 + i_2 \qquad (5.13)$$

where $A = 9\, q_0 \pi^2 V^2 M^2 / 2R^2 \lambda^4$, and V is the volume of the particle, λ is the wavelength of light, M is a function of the refractive index (n) of the particle material $[M = (n^2 - 1)/(n^2 + 2)]$ and R is the distance from the scattering particle.

The way in which the amount of scattered radiation varies with direction is shown in Figure 5.5.

The upper limit of size for the Rayleigh law (Equation 5.13) is about 0.1λ but the lower limit extends down to molecular dimensions.

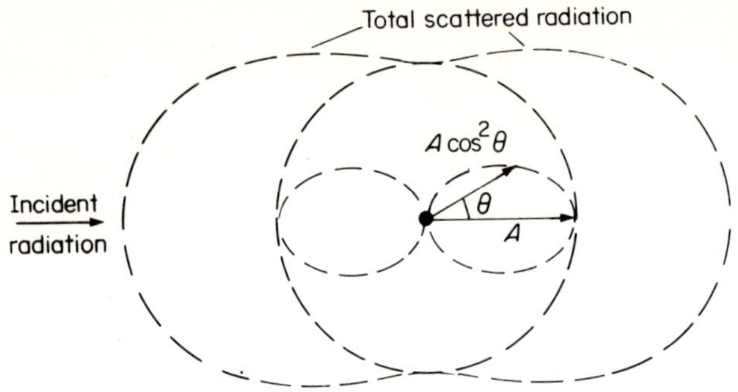

Figure 5.5 Radiation scattered by a small particle

Integration of Equation 5.13 over 2π radians gives the total intensity of scattered radiation:

$$S = 24\ \pi^3\ M^2\ V^2/\lambda^4 \qquad (5.14)$$

F_{sc}, the ratio of the apparent scattering cross-sectional area to the geometrical cross-sectional area, equals $4S/\pi d^2$.

Expressing S in terms of the particle diameter:

$$F_{sc} = 8\ \pi^4\ d^4\ M^2/3\lambda^4$$

$$= 8\ x^4\ M^2/3 \qquad (5.15)$$

where $x = \pi d/\lambda$. The attenuation cross-section equals the sum of the cross-sections of scattering and absorption. In the case of a dielectric sphere, absorption is zero and

$$F_{at} = F_{sc} = 8\ x^4\ M^2/3 \qquad (5.16)$$

For an absorbing sphere, the expressions for scattering and absorption, and, therefore, attenuation, involve the complex refractive index $n' = n(1 - ik)$, where i is the square root of -1 and k is the absorption index. M becomes $M' = (n'^2 - 1)/(n'^2 + 2)$, and

$$F_{at} = F_{sc} + F_{ab}$$

The scattering and absorption cross-sections have been shown to be equal to $8\, x^4\, M'^2/3$ and $4\, x\, \mathrm{Im}(M')$ respectively where $\mathrm{Im}(M')$ is the imaginary part of M'. Therefore,

$$F_{at} = 8\, x^4\, M'^2 + 4\, x\, \mathrm{Im}(M') \qquad (5.17)$$

Equation 5.17 reduces to the Rayleigh expression for a non absorbing sphere if k, the absorption index, is set equal to zero.Substituting $n' = n(1 - ik)$, into each part of Equation 5.17 gives:

$$F_{ab} = \frac{24\, n^2\, kx}{[n^2(1 - k^2) + 2]^2 + 4\, n^4\, k^2} \qquad (5.18)$$

and

$$F_{sc} = \frac{8\, x^4\{[n^2(1-k^2)-1][n^2(1-k^2)+2]+4\, n^4\, k^2\}^2+36\, n^4\, k^2}{3\{[n^2(1-k^2)+2]^2+4\, n^4\, k^2\}^2}$$

(5.19)

Soot has a refractive index of $n = 1.5$ to 3 and an absorption index $k = 0.1$ to 1 with higher values corresponding to those soots of high carbon content. Soot appears to consist of agglomerates or chains of approximately spherical particles. However, if one can assume that agglomeration does not seriously modify the radiative properties of the individual particles and that the particles possess diameters in the range 0.02 to 0.06 μm, then the attenuation cross-section of the soot is given by Equation 5.17.

Evaluation of F_{ab} and F_{sc} shows F_{sc} to be negligible in comparison to F_{ab} which, from Equation 5.18, is:

$$F_{ab} = F_{at} + 24\, \pi d n^2 k/[\lambda(n^2 + n^2k^2)^2 - 4(n^2 - n^2k^2 + 1)]$$

$$= 24\, \pi d \phi/\lambda \qquad (5.20)$$

where $\phi = n^2k/[(n^2 + n^2 k^2)^2 + 4(n^2 - n^2 k^2 + 1)]$, and, if n and k are constant with wavelength, then

$$F_{ab} \propto d \tag{5.21}$$

and absorption by a particle is proportional to d^3, that is, its mass.

Since F_{ab} is inversely proportional to $1/\lambda$, a soot particle is not grey, and radiation absorbed, or emitted, by soot particles must be considered at single wavelengths. The monochromatic absorption coefficient (K_λ) of a soot suspension is $F_{ab}\ c_v\ S_v/4$ or, expressing $S_v = 6/d$ and F_{ab} as in Equation 5.20,

$$K_\lambda = 36\ \pi\ c_v\ \phi/\lambda = \kappa\ c_v/\lambda \tag{5.22}$$

$$= \kappa\ c_m/\lambda\ \rho \tag{5.23}$$

where κ is $36\ \pi\ \phi$, and c_v and c_m are the concentrations of particles in the flame, by volume and mass respectively.

The monochromatic emissivity is:

$$\varepsilon_\lambda = 1 - \exp(-K_\lambda\ \ell) \tag{5.24}$$

Thus, the monochromatic emissivity of a flame of mean beam length 1 m containing soot at a concentration of $10^{-7}\ m^3\ m^{-3}$ is shown in Figure 5.6. For this calculation, based on Equation 5.22, the optical properties of the soot were assumed to be: $n = 1.5$ and $k = 0.6$.

Equation 5.22 shows that the absorption coefficient is independent of particle size (as long as the particles are less than 0.05 μm) and is proportional to the mass concentration of the soot in the flame.

The average emissivity is obtained from Equation 4.15:

$$\varepsilon = \int_0^\infty \varepsilon_\lambda\ E_{b\lambda}\ \Big/ \int_0^\infty E_{b\lambda} \tag{5.25}$$

$$= \int_0^\infty [1 - \exp(-K_\lambda\ \ell)]\ E_{b\lambda}\ \Big/ \int_0^\infty E_{b\lambda} \tag{5.26}$$

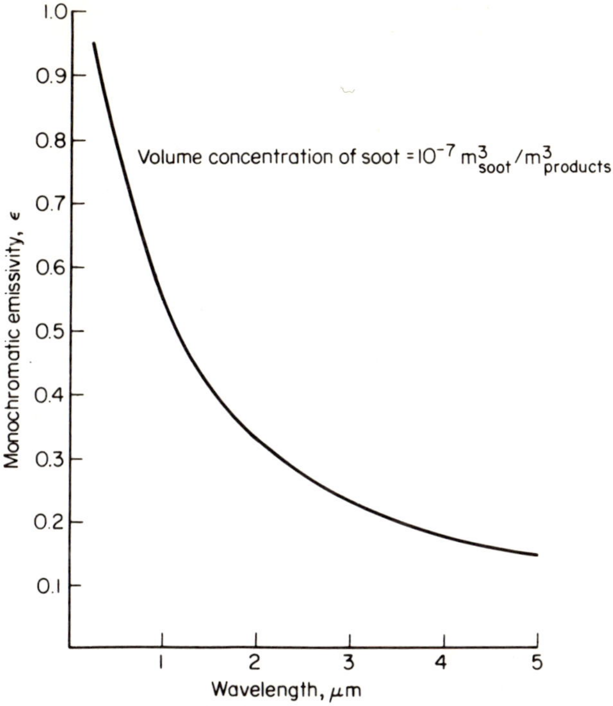

Figure 5.6 Emissivity of a cloud of soot particles

If n and k are assumed to be independent of wavelength, substitution of the expression for K_λ in Equation 5.26 and appropriate cancellation gives:

$$\varepsilon = 1 - \left[1/(1 + \kappa\, c_v\, L\, T/c_2)^4\right] \qquad (5.27)$$

Using the same values in Equation 5.27 as those used in the calculation of the monochromatic emissivity of a cloud of soot particles (Figure 5.6) gives an average emissivity at a temperature of 1600 K of 0.29. From studies on coal, κ varies from 7.5 to 3.7 as the atomic ratio H/C varies from 0 to 0.4 (McCartney and Ergun, 1958). Studies on oil flames at the Flame Radiation Research Centre at Ijmuiden produced values of 6.3 and 14.1 (Thring, 1962).

5.2.3 Intermediate Sizes

When the wavelength of the radiation is comparable to the particle size, the field cannot be considered uniform throughout the particle and electrons in various locations within the particle can be accelerated in different directions. Equally, geometric optics, applicable to large particles, cannot be used because diffraction cannot be distinguished from the scattering that occurs within the bulk of the particle. Mie solved the problem precisely for spherical particles and solutions now exist for particles of other shapes.

Mie theory

Corresponding to Equation 5.13, the Mie equation is:

$$I_\theta = i_1 + i_2 = \lambda^2(i_1' + i_2')/8\,\pi^2 R^2 \qquad (5.28)$$

i_1' and i_2' are known as the intensity functions and are defined in terms of the coefficients of the electric and magnetic waves a_n and b_n respectively. a_n and b_n are complex functions of particle diameter, refractive index and wavelength of radiation.

The total scattering is given as:

$$S = \lambda^2/2\pi \sum_{n=1}^{\infty} (2n + 1)(|a_n|^2 + |b_n|^2) \qquad (5.29)$$

The extinction is also of similar form:

$$E = \lambda^2/2\pi \sum_{n=1}^{\infty} (2n + 1)\left[\mathrm{Re}(a_n + b_n)\right] \qquad (5.30)$$

where Re refers to the real part of the function.

The application of this kind of theory to particles in flames or combustion products is largely confined to research work. In practice, char particles are assumed to be large convex black bodies (Section 5.2.1.), soot is assumed to behave in the manner described in Section 5.2.2 and ash particles are largely ignored because data on the radiative properties are not available.

5.3 REPRESENTING REAL GASES

In the calculation of heat transfer in enclosures containing absorbing media, it is often assumed that the media are grey. For a grey medium, the attenuation of intensity of a beam of radiation is exponential with distance, and its emissivity is:

$$\varepsilon = 1 - \exp(-k\ p\ \ell) \qquad (5.30)$$

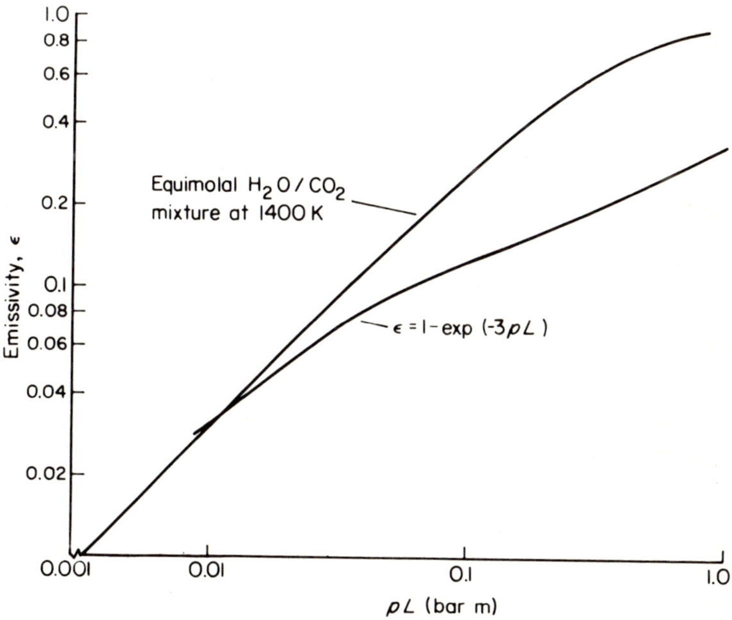

Figure 5.7 Emissivities of real and grey gases

However, because of their spectral absorbing characteristics, real gases do not show this exponential behaviour (Figure 5.7). Thus, although the assumption of greyness may be adequate for a simple calculation, the departure from greyness must be taken into account when greater accuracy is required. Although calculation techniques, such as the Monte Carlo Method, can cater for spectral variations, it is possible to make an analysis without using the full details of the spectral variations. The structures of the leaving flux equations are such that it is convenient to retain the concept of greyness, and this is done by equating the emissivity of the real gas with a weighted sum of the emissivities of a number of hypothetical grey gases:

$$\varepsilon_g = a_1[1 - \exp(-k_1pL)] + a_2[1 - \exp(-k_2pL)] + a_3[1 - \exp(-k_3pL)] + \ldots$$

$$= \sum a_i - \sum a_i \exp(-k_ipL) \qquad (5.31)$$

where the weighting coefficients a_i and the absorption coefficients k_i are determined by means of a curve fitting technique. A graphical technique which allows this to be done is as follows.

A plot of ε_g versus pL is drawn as in Figure 5.7. At large pL, $\exp(-k\ pL)$ is small, and

$$\sum a_i = \varepsilon_g \qquad (5.32)$$

In order to determine the individual weighting coefficients and the absorption coefficients, Equation 5.31 is rewritten in the form:

$$\sum a_i - \varepsilon_g = \sum a_i \exp(-k_ipL) \qquad (5.33)$$

If it is now assumed that $k_1 < k_2 < k_3 < \ldots$, then, at large pL,

$$\exp(-k_1pL) >> \exp(k_2pL) >> \exp(-k_3pL) \qquad (5.34)$$

and Equation 5.33 becomes:

$$\sum a_i - \varepsilon_g = \exp(-k_1pL) \sum a_i$$

and $$\ln(\sum a_i - \varepsilon_g) = -k_1pL + \ln(\sum a_i) \qquad (5.35)$$

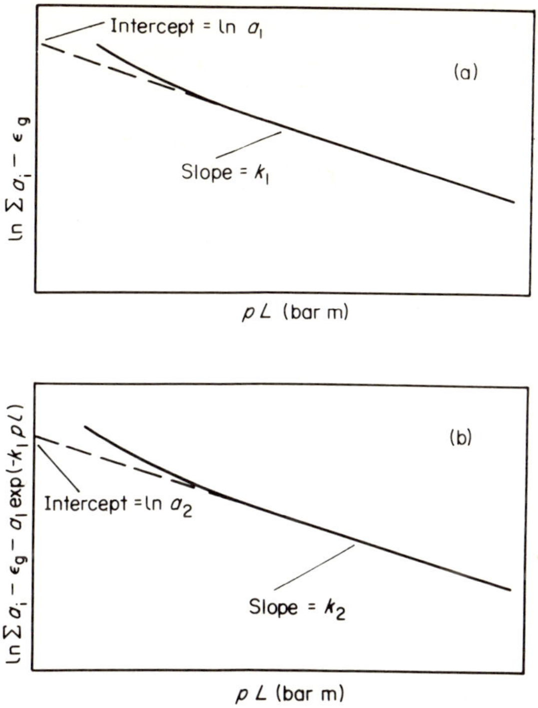

Figure 5.8 Coefficients of a grey gas fit

If $\ln(\Sigma a_1 - \varepsilon_g)$ is plotted against pL as in Figure 5.8a, then $\ln(a_1)$ and k_1 can be determined by drawing the tangent to the curve at large pL, and measuring its gradient (k_1) and its intercept [$\ln(a_1)$] on the ordinate. a_2 and k_2 are obtained in a similar way by plotting $\ln[\Sigma a_i - \varepsilon_g - a_1 \exp(-k_1 pL)]$ against pL, (Figure 5.8b). The number of terms required to produce a good fit depends on the range of pL over which representation is required and upon the shape of the $\ln(\Sigma a_1 - \varepsilon_g)$ versus pL curve for the real gas. The more the shape deviates from exponential, the more terms are required. It is, however, rarely necessary to determine more than three or four terms to provide an adequate fit.

The procedure can be repeated at different temperatures to allow the variation of the coefficients with temperature to be determined. In practice, the values of k are assumed to be temperature independent, and temperature dependence is described solely by the weighting coefficients.

The absorptivity of a real gas may also be represented by the absorptivities of a number of grey gases, but, in this case, the coefficients are functions of the temperature of the source of radiation as well as of gas temperature (Sarofim, 1962).

6 Heat Transfer from Flames

Heat transfer from flames takes place by convection and radiation; conduction, except in its incorporation in convective heat transfer, is unimportant. But, to paraphrase Hottel and Sarofim (1967): 'it is inappropriate to make very detailed calculations for an engineering process if simpler ones provide answers which are just as useful'. Thus, it is often useful to establish whether one needs to take both radiation and convection into account. The following estimates of heat transfer in a boiler emphasises this point.

6.1 HEAT TRANSFER IN THE COMBUSTION CHAMBER OF A BOILER

6.1.1 Radiative Heat Transfer

For pulverised fuel with a weight mean size of 50 μm and a Rosin Rammler size distribution with $n = 1.2$, the specific surface is about 200 $m^2\ kg^{-1}$. When dispersed in air at 1600 K, the mass concentration is about 1.5×10^{-2} kg m^{-3}. From Equation 4.11 ($K = c_m S_m/4$), the absorption coefficient is 0.75 m^{-1}. Using a mean beam length of 0.6 m for a cylindrical combustion chamber of diameter 1 m, the emissivity is:

$$\varepsilon = 1 - \exp(-0.75 \times 0.6) = 0.362$$

For combustion products at 1600 K, the rate of heat radiated to the walls will be:

$$q = 0.362 \times 56.7\ (1.6)^4 \simeq 135 \text{ kW m}^{-2}$$

This value is a rough guide to what might be expected in the combustion chamber of a boiler in which the walls are not highly reflecting and the temperatures of the walls are relatively low.

6.1.2 Convective Heat Transfer in a Boiler

(a) *Water tube boiler*

Field *et al.* (1967) have suggested that, on the basis that a thick stagnant layer of gas exists in the vicinity of the walls of a large water tube boiler, heat transfer to the walls is by natural convection, that is, apart from radiation.

The following data allow the Grashof and Prandtl numbers to be calculated and the appropriate correlation chosen. The property values of nitrogen (Table 3.5), in preference to those of the combustion products, have been used in this calculation because greater precision is not required.

Data

Height of wall, L	8 m
Temperature of stagnant layer, T_g	1400 K
Temperature of wall, T_W	600 K
Mean temperature = (1400 + 600)/2	= 1000 K (the temperature used for the evaluation of physical properties)
Thermal conductivity of gases, k	6.65×10^{-2} W m^{-1} K^{-1}
Density of combustion gases	0.37 kg m^{-3}
Coefficient of cubical expansion	1×10^{-3} K^{-1}
Viscosity of combustion products	3.97×10^{-5} N s m^{-2}
Specific heat of combustion products	1.17×10^{3} J kg^{-1} K^{-1}

$$Gr = \frac{8^3 \times (0.37)^2 \times 9.81 \times 10^{-3} \times 800}{(3.97 \times 10^{-5})^2} = 3.49 \times 10^{11}$$

$$Pr = \frac{1.17 \times 10^3 \times 3.97 \times 10^{-5}}{6.65 \times 10^{-2}} = 0.7$$

Since the Grashof number is greater than 10^9, Equation 2.36 give the appropriate correlation.

$$Nu = 0.1\ (Gr\ Pr)^{0.33} = 572$$

$$h_c = (Nu)k/L = 572 \times 6.65 \times 10^{-2}/(8 \times 1000) = 4.75 \times 10^{-3}\ \text{kW m}^{-2}\ \text{K}^{-1}$$

and

$$q = h_c\ \Delta T = 3.8\ \text{kW m}^{-2}$$

(b) Shell boiler

The flow of combustion products in a shell type boiler is likely to be turbulent as the following data show. Property values of nitrogen (Table 3.5) have again been used in this order-of-magnitude calculation.

Data

Diameter and length of combustion chamber, $D = 1$ m, $L = 6$ m
Velocity of fluid, $U = 8\ \text{m s}^{-1}$
Fluid temperature in the main stream, $T_g = 1400$ K
Wall temperature, $T_w = 600$ K
Mean film temperature, $T_f = 1000$ K

Values of the thermal conductivity, density, viscosity and specific heat of the combustion products are the same as those used in the calculation for the water tube boiler.

$$Re = D\ u\ \rho/\mu = \frac{1 \times 8 \times 0.37}{3.97 \times 10^{-5}} = 7.46 \times 10^4$$

$$Pr = c_p \mu/k = 0.7 \text{ (as for water tube boiler above)}$$

The correlation for heat transfer in a long tube of circular cross-section (Equation 2.29) must be modified in the following way for short tubes with $2 < L/D < 20$ (Kreith, 1973):

$$Nu = 0.023\, Re^{0.8}\, Pr^{0.33} \left[1 + (D/L)^{0.7}\right]$$

$$\frac{h_c D}{k} = 0.023 \times (7.46 \times 10^4)^{0.8} \times 0.7^{0.33} \times \left[1 + (1/6)^{0.7}\right]$$

$$= 0.023 \times 7.91 \times 10^3 \times 0.889 \times 1.29$$

$$= 209$$

$$\therefore h_c = 209 \times 6.65 \times 10^{-2}$$

$$= 13.9 \text{ W m}^{-2} \text{ K}^{-1}$$

and

$$q = h_c \; T = 13.9 \times 10^{-3} \times 800$$

$$= 11.1 \text{ kW m}^{-2}$$

Consequently, these approximate calculations of the relative contributions of convective and radiative heat transfer in a boiler show that the convective component is comparitively small compared to that of radiation. Thus, it is often sufficient to make relatively simple estimations of convective heat transfer or, in some cases, to neglect it.

6.2 MODELS OF HEAT TRANSFER FROM A FLAME TO ITS ENCLOSURE

In most applications, the flame and combustion products are enclosed and calculations usually involve an estimate of the heat transfer from the flame to either a part or the whole of the enclosure. The mathematical analysis of a practical problem of this kind is often called mathematical modelling, and this chapter will be concerned with two models, the well-stirred model and the long chamber model.

6.2.1 The Well-Stirred Model

This model is characterised by the assumption that there are no temperature or concentration gradients throughout the chamber. In practice, an approximation to this occurs when combustion takes place prior to entry to, or in a very small part of, the chamber, coupled with a sufficiently high momentum of the incoming gases to ensure that the combustion products are well-stirred. Example are oil preheating furnaces and rapid reheating furnaces in the metallurgical industries. The combustion products in the chamber can be considered as a single gas zone of uniform temperature T_g. The problem is to determine T_g, the rate of heat transfer to the enclosure and the efficiency of heat transfer for a given fuel/air ratio and fuel flow rate.

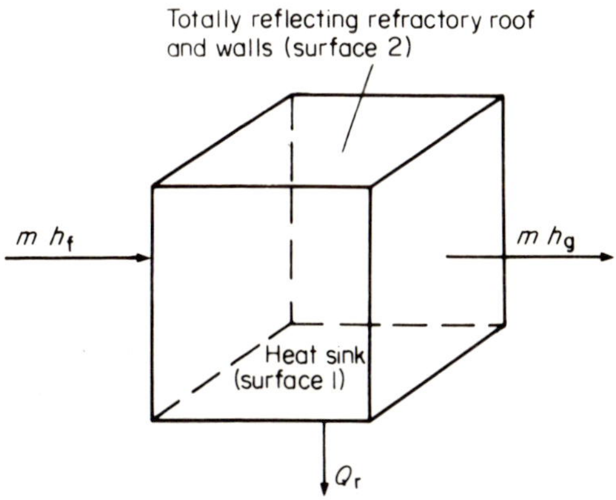

Figure 6.1 Cubical combustion chamber showing thermal balance

As an example, consider a cubical enclosure (Figure 6.1) with a heat sink covering one face (surface 1) and the other five faces (surface 2) comprising refractory. Considering heat transfer by radiation alone, the net heat transfer from the flame to the heat transfer surface may be expressed as:

$$Q_r = A_1 \varepsilon_1 (W_1 - \sigma T_1^4)/\rho_1 \tag{6.1}$$

where W_1 is determined from the solution of the two equations:

$$W_1 = \varepsilon \sigma T_1^4 + \rho(W_2 F_{12} \tau_{12} + \varepsilon_{g1} \sigma T_g^{4}) \tag{6.2}$$

$$W_2 = W_1 F_{21} \tau_{12} + W_2 F_{22} \tau_{22} + \varepsilon_{g2} \sigma T^{4} \tag{6.3}$$

in which everything is known apart from W_1 and W_2. Equation 6.3 implies that the refractory walls re-radiate all the radiation incident upon them. For a cubical enclosure,

$$F_{12} = 1; \quad F_{21} = 0.2; \quad F_{22} = 0.8$$

assuming that $\varepsilon_{g1} = \varepsilon_{g2} = \varepsilon_g$ and $\tau_{12} = \tau_{22} = \tau = 1 - \varepsilon_g$.

The leaving flux from surface 1 (W_1) may be derived from Equation 6.2 and 6.3 and expressed in terms of E_1 ($\varepsilon_1 \sigma T_1^4$) and E_g ($\varepsilon_g \sigma T_g^{4}$) thus:

$$W_1 = \frac{E_1 + \rho_1 E_g \left[1 + \tau/(1 - 0.8\tau)\right]}{1 - 0.2 \rho \tau^2/(1 - 0.8\tau)} \tag{6.4}$$

Substitution of Equation 6.4 into Equation 6.1 enables Q_r to be expressed in terms of the gas temperature T_g. The gas temperature will be less than the adiabatic flame temperature T_f because of heat transfer to the heat sink, and the net heat gained by the sink is equal to the difference in enthalpy between the air and gas entering the chamber and that of the combustion products leaving.

$$Q_r = m(h_f - h_g) = m\overline{C}_p(T_f - T_g) \tag{6.5}$$

where m is the mass flow rate of products, h_f is the enthalpy per unit mass of products at the adiabatic flame temperature T_f and h_g is the enthalpy per unit mass of products at the temperature T_g and $\overline{C}_p$ denotes a mean specific heat appropriate for the temperature difference $(T_f - T_g)$.

The composition of the products of combustion and the adiabatic flame temperature can be calculated (Appendices 1 and 2) and the emissivity of the products determined (Section 5.1) for any air/fuel ratio. Equation 6.5 may now be solved for T_g, given a heat sink temperature.

It may be convenient to use a numerical method of solution or possibly a graphical method as illustrated in Figure 6.2.

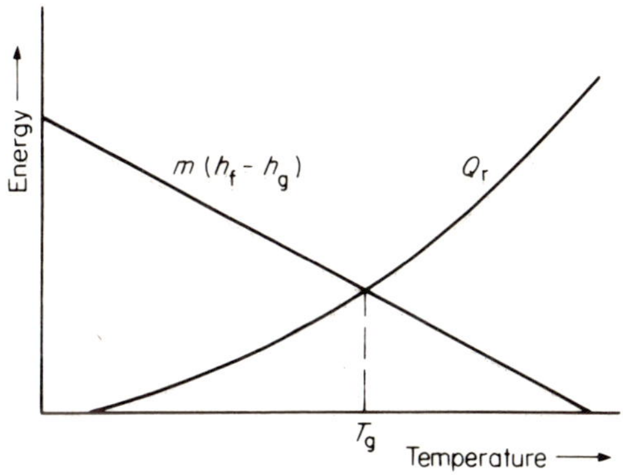

Figure 6.2 Graphical solution of Equation 6.5

Q_r can be evaluated and the efficiency, η, of the furnace or chamber can conveniently be expressed as the ratio of heat transferred to available enthalpy.

$$\eta = Q_r/m\,h_f \qquad (6.6)$$

In addition, the firing density, *D*, of the chamber can be defined as the heat available per unit area of heat sink:

$$D = m\,h_f/A \qquad (6.7)$$

The variation of the efficiency of heat transfer with firing density can be plotted as in Figure 6.3. The approximate operating ranges of a number of combustion systems have been indicated, and it is clear that efficiencies can be increased by

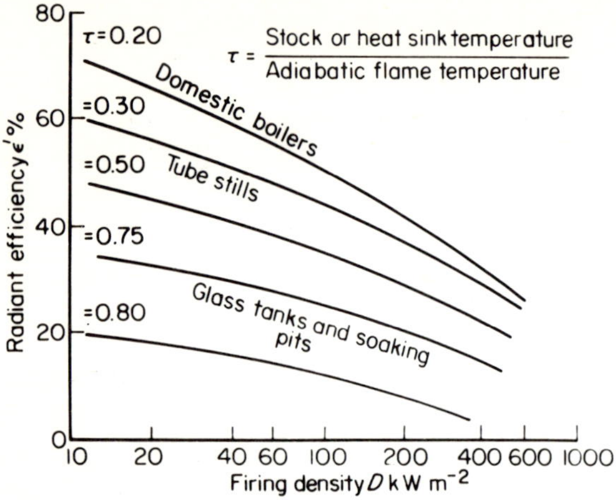

Figure 6.3 *Approximate efficiencies of different combustion systems*

reducing the firing density.

If convection is significant, a convective term can be added to the left hand side of Equation 6.5:

$$m(h_f - h_g) = Q_r + h\,A\,(T_g - T_1) = Q_t \qquad (6.8)$$

and the efficiency is redefined as:

$$\eta = Q_t / m\,h_f \qquad (6.9)$$

In furnaces which cannot be considered so well-stirred, the gases leave the furnace at a temperature lower than the gas radiating temperature. This effect can be catered for by writing the enthalpy of the gases leaving the furnace as $m\,c_p\,(T - \Delta)$ where Δ is the temperature difference between the temperature of the products in the chamber and that of the products leaving. Efficiencies calculated using the 'not-so-well-stirred' model are greater than efficiencies calculated using the well-stirred model because of the greater use of available energy. Consequently, the use of a well-stirred model may well lead to over design.

Hottel and Sarofim (1967) have used a model of this type to predict heat transfer rates in the radiant section of tube stills and have obtained correlations between prediction and experimental data within an average of 4%. Davies and Oeppen (1972) used a similar model to compare heat transfer rates between heavy fuel oil flames and natural gas flames. Their results indicated that, in some circumstances, higher heat transfer rates can be obtained with gas flames than with oil flames, in spite of the higher emissivity of the latter.

The extension of this type of analysis to non-steady state systems (such as, the heating of a billet in a rapid-heating furnace) is relatively straightforward (Davies, Lucas and Toth, 1971). In this application, appropriate data on convective heat transfer coefficients were not available and the naphthalene sublimation technique (Section 2.5.2) was used to provide suitable data.

6.2.2 The Long Chamber Model

If the combustion chamber is long compared with its cross-sectional dimensions, substantial temperature gradients will occur within the chamber. In these circumstances, it may be appropriate to use the 'long chamber model'. Combustion is assumed to occur at one end of the furnace in a well-stirred region and the products are then assumed to flow along the furnace with decreasing temperature, but with no radial variation in temperature. This type of flow is commonly described as 'plug flow' and the heat transfer calculation can most conveniently be done by dividing conceptually the plug flow region into a number of sections or zones, each characterised by constancy of temperature (Figure 6.4). The temperature distribution along the furnace is therefore approximated by a step function, and the shorter the section lengths and the greater the number of sections, the better the approximation to the true temperature distribution.

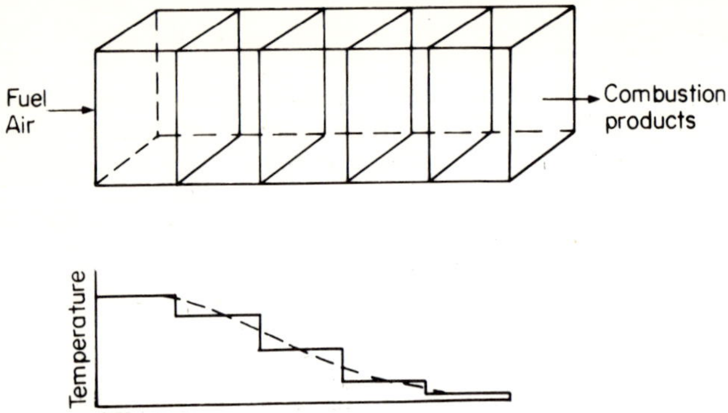

Figure 6.4 *Approximation of a continuous temperature distribution by a step function*

In many cases, it is convenient to use sections of differing lengths, with short sections in regions of high temperature gradient, and larger sections in regions of low temperature gradient. The analysis of heat transfer in this type of model may be carried out by more than one method. The simplest analysis involves assuming that each section receives no radiation from neighbouring sections, and that each section can be treated as a well-stirred zone. The equation for each zone is:

$$Q_r + Q_c + m(h_i - h_o) = Q_{comb} \qquad (6.10)$$

where h_i and h_o are the enthalpies at the inlet and outlet of the zones, corresponding to inlet and outlet temperatures, T_i and T_o; Q_r and Q_c are evaluated at a mean gas temperature $(T_i + T_o)/2$ and Q_{comb} is the heat released by combustion in the zone.

The inlet temperature of each zone is equal to the outlet temperature of the previous zone, and if the calculations are carried out from the combustion zone in the direction of flow, then outlet temperatures can be determined for each zone in turn.

This analysis does not take into account radiative exchange between different sections of the furnace, although there is no reason why this should not be done when formulating the equations for each section. In practice, however, the exchange between different sections is often neglected because of two difficulties. Firstly, there is the problem of determining the view-factors and transmissivities between sections that are not adjacent. Secondly, the energy balance equations for each individual section involve the unknown gas temperatures in all the other sections and not merely the gas temperature for the preceding section. Consequently, the energy balance equations, which are not linear because of the radiative terms, have to be solved simultaneously.

An example of the application of the long chamber model is in the prediction of gas temperatures in a p.f.-fired water tube boiler (Field *et al.*, 1967). Figure 6.5 is a representation of the combustion chamber. The flame and combustion products are assumed to occupy a volume 20 m long with a square cross-section 4 m × 4 m. This volume is divided into five zones and combustion is assumed to be completed in the first one.

Other data:

Walls of chamber	: black at 750 K
Absorption coefficient of combustion products throughout volume	: 0.2 m^{-1}
Interfaces between zones	: radiatively reflecting surfaces (no interchange of radiation between adjacent zones)
Temperature of fuel air stream at inlet	: 500 K
Total flow rate	: 100 kg s^{-1}
Specific heat of combustion products	: 1.26 kJ kg^{-1} K^{-1}
Heat generated on combustion	: 2×10^5 kW

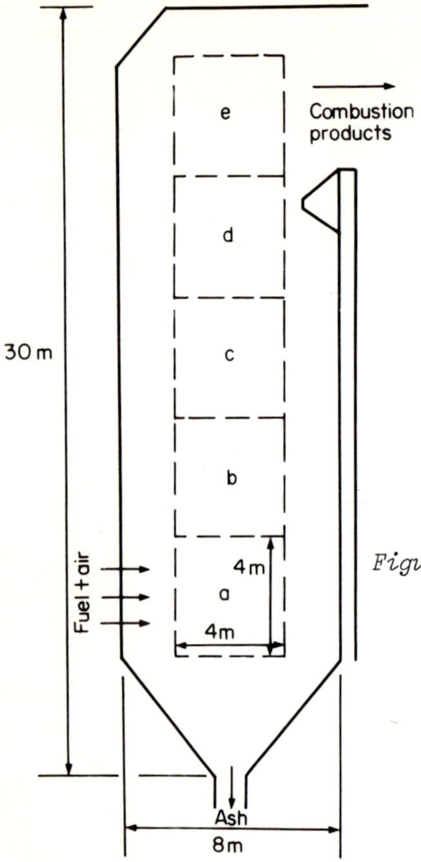

Figure 6.5 Diagram of the combustion chamber of a p.f.-fired water tube boiler

The temperature distribution throughout the boiler and the heat transferred may be determined from a heat balance on each zone (Equation 6.10).

(a) Radiation

Each zone can be considered to be bounded by a black surface (1) and a reflecting surface (2). In the case of the zones b, c and d, the two reflecting planes are treated as a single surface. Using the leaving flux notation, the net radiative heat transfer to the black surface is:

$$Q_{r1} = A_1 (H_1 - W_1) = A_1 (H_1 - E_{b1})$$
$$= A_1 (E_{b1} F_{11} \tau_{11} + W_2 F_{12} \tau_{12} + \varepsilon_{g1} E_{bg} - E_{b1}) \quad (6.11)$$

where W_2 is given by:

$$W_2 = W_2 F_{22} \tau_{22} + E_{b1} F_{21} \tau_{21} + \varepsilon_{g2} E_{bg} \tag{6.12}$$

The mean-beam lengths used in calculating the transmissivities and emissivities were derived from Dunkle's data, using the procedure described by Gray and Müller (1974).

Zones a and e

$A_1 = 80 \text{ m}^2$

$F_{11} = 0.8; \quad F_{12} = 0.2; \quad F_{22} = 0; \quad F_{21} = 1$

$\tau_{11} = 0.597 \qquad \tau_{12} = 0.6$

$\varepsilon_{g_1} = \varepsilon_{g_2} = 0.38$

Substituting these values in Equation 6.12 gives

$W_2 = 0.6\, E_{b_1} + 0.38\, E_{bg}$

and

$Q_{r1} = 2050\, (T_g/1000)^4 - 645 \text{ kW}$

Zone b c d

$A_1 = 64 \text{ m}^2$

$F_{11} = 0.6 \quad F_{12} = 0.4 \quad F_{22} = 0.2 \quad F_{21} = 0.8$

$\tau_{11} = 0.561; \quad \tau_{12} = 0.654; \quad \tau_{22} = 0.413$

$\varepsilon_{g_1} = \varepsilon_{g_2} = 0.38$

Substituting these values in Equation 6.12 gives:

$W_2 = 0.57\, E_{b_1} + 0.414\, E_{bg} \text{ kW}$

and

$Q_{r_1} = 1770\, (T_g/1000)^4 - 590 \text{ kW}$

(b) Convection

Heat transfer by natural convection to the vertical walls of a water tube boiler is (see Section 6.1.2.)

$$q_c = 4.75 \times 10^{-3} \; (T_g - T_w) \quad \text{kW m}^{-2}$$

where T_g and T_w are the gas and wall temperatures respectively. For each zone:

$$Q_c = 64 \times 4.75 \times 10^{-3} \; (T_g - 750) \text{ kW}$$

$$= 0.304 \; (T_g - 750) \text{ kW}$$

Heat balance on zone a

From Equation 6.10:

$$2050(T_g/1000)^4 - 645 + 0.304(T_g - 750) +$$

$$100 \times 1.26(T_a - 500) = 2.0 \times 10^5 \qquad (6.13)$$

where T_g is the mean gas temperature in zone a. In this zone, the temperature is assumed constant and equal to T_a, the gas temperature at the outlet of zone a. Equation 6.13 allows the outlet temperature (the inlet temperature of the next zone) to be determined.

Heat balance on zone b

$$1770(T_g/1000)^4 - 590 + 0.304(T_g - 750) +$$

$$126 \; (T_b - T_a) = 0 \qquad (6.14)$$

where T_g is assumed to be the mean of the inlet and outlet temperatures respectively $(T_a + T_b)/2$ and T_b is the gas temperature at the outlet of zone b.

Similar balances may be drawn up for the other zones and the outlet temperature of each zone determined in turn.

Table 6.1 shows the calculated values of temperatures and heat transferred by radiation and convection for each of the zones.

Table 6.1. Heat transfer in a water tube boiler

Zone	Inlet and Outlet temperature (K)		Heat transferred (kW) Radiation	Convection
a	500	1885	2.588×10^4	3.6×10^2
b	1885	1737	1.902×10^4	3.2×10^2
c	1737	1627	1.415×10^4	2.8×10^2
d	1627	1541	1.114×10^4	2.5×10^2
e	1541	1469	0.955×10^4	2.2×10^2

The extension of the simplest treatment of the long furnace model is to make some estimate of the heat transferred by radiation from one zone to another. This procedure will serve to lower the estimated exit temperature from the first zone in Table 6.1 and raise the estimated temperatures of the combustion products in later zones. A comparitively simple way of estimating the heat transferred from one zone to another would be to use the concept of an imaginary surface between the zones, for which the following statement relating to energy arriving and leaving the surface is made:

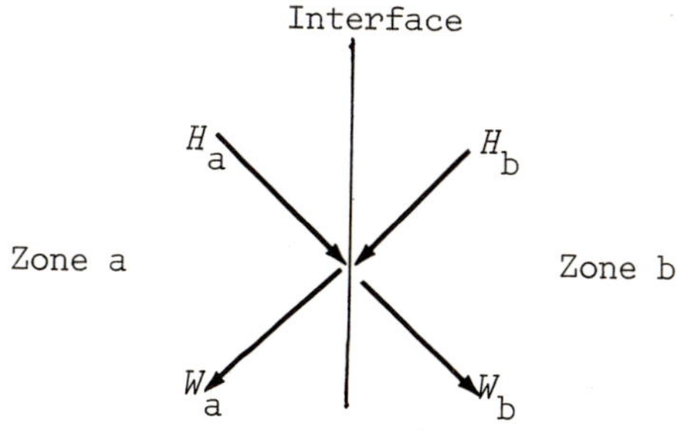

$$H_a = W_b \text{ and } H_b = W_a$$

where *H* and *W* are the incident and leaving fluxes respectively. The technique is that used by Kolenda and Gray (1974) to asses heat transfer in rotary kilns and is described in the text by Gray and Müller (1974).

6.3 MORE COMPLEX MODELS

Greater accuracy in predicting furnace performance may be achieved, provided that appropriate data on physical properties are available, if the furnace chamber is divided into a sufficient number of zones, both axially and longitudinally, such that the statement that the physical properties throughout each zone are constant is a near approximation to the truth. Although, in principle, the technique is similar to the simple zoning technique already described, the calculation increases considerably in complexity as the number of zones and bounding surfaces involved increases. Such a calculation has been described by Hottel and Sarofim (1967) for a relatively simple furnace of square cross-section.

Appendix 1
Composition of Products of Combustion

The equilibrium composition of the products of combustion may be calculated relatively easily, although the amount of computation increases as more dissociated species are taken into account. The simplest calculation is based on the assumption of no dissociation, an assumption which is acceptable at low temperatures (below 1600 K). Such a calculation also provides a first estimate of the concentration of some components even when dissociation is considered.

The following calculations involve the combustion of the hydrocarbon $C_u H_v$ in air. If the molecular formula of the fuel is not known, then the calculation is based on 100 kg of the fuel and an empirical formula derived as follows:

100 kg coal

Constituent	*Weight* (kg)	*Weight* / *Atomic or molecular weight*
Carbon	74.0 % wt	74/12 = 6.17
Hydrogen	5.5 % wt	55/1 = 5.5
Oxygen	8.6 % wt	8.6/16 = 0.538
Sulphur	1.2 % wt	1.2/32 = 0.038
Nitrogen	1.0 % wt	1.0/14 = 0.072
Water	5.0 % wt	5.0/18 = 0.278

and the empirical formula is:

$$C_{6.17}\ H_{5.5}\ O_{0.538}\ S_{0.038}\ N_{0.072}\ (H_2O)_{0.278}$$

(1) COMBUSTION WITH NO DISSOCIATION

Combustion in air is represented by:

$$C_u\ H_v + z\ O_2 + 3.76\ z\ N_2 \rightarrow u\ CO_2 + (v/2)H_2O + e\ O_2 + 3.76\ z\ N_2 \qquad (A1)$$

where $z = u + v/4 + e$; e = number of moles of excess oxygen (or air) and 3.76 is the ratio by volume of nitrogen to oxygen in the air. The total number of moles of products, N, is

$$N = u + v/2 + e + 3.76z$$

and the mole fractions and partial pressures of each component are given in Table A1.1.

Table A1.1. Partial pressures of combustion products

Component	Mole fraction	Partial pressure*
CO_2	u/N	uP/N
H_2O	$v/2N$	$vP/2N$
O_2	e/N	eP/N
N_2	$3.76\ z/N$	$3.76zP/N$

*P is the combustion pressure.

At flame temperatures resulting from the combustion of hydrocarbon fuels (approximately 2000°C if no heat loss from the flame occurs) the only dissociated species which are usually taken into account in the calculation of flame temperatures are carbon monoxide and hydrogen which can be considered to result from the dissociation of carbon dioxide and water.

(2) DISSOCIATION OF CARBON DIOXIDE AND WATER

A simple iterative procedure which is relatively insensitive to the initial estimate on which the iteration is based has been suggested by Mitchell (1976). It is a variant of the technique discussed by Harker (1967) in that an estimate of the partial pressure of only one component is required.

Consider 1 kilomole of dissociated products at a pressure of P bar:

$$p_{CO} + p_{CO_2} + p_{O_2} + p_{H_2} + p_{H_2O} + p_{N_2} = P \qquad \text{(A2)}$$

The equilibrium coefficients used to describe the dissociation are:

$$K_1 = \frac{p_{CO}\, p_{O_2}^{\frac{1}{2}}}{p_{CO_2}} \qquad \text{(A3)}$$

and

$$K_2 = \frac{p_{H_2}\, p_{O_2}^{\frac{1}{2}}}{p_{H_2O}} \qquad \text{(A4)}$$

All the partial pressures are expressed in terms of the atomic ratios, the number of kiloatoms of oxygen in 1 kilomole of combustion gases (n_O) and the partial pressure of oxygen (p_{O_2}).

$$k = n_C/n_O \qquad \text{(A5)}$$

$$\ell = n_H/n_O \qquad \text{(A6)}$$

$$m = n_N/n_O \qquad \text{(A7)}$$

The number of katoms of oxygen can be expressed in terms of the partial pressures of other constituents by:

$$P\,k\,n_0 = p_{CO} + p_{CO_2} \tag{A8}$$

$$P\,\ell\,n_0 = 2p_{H_2} + 2p_{H_2O} \tag{A9}$$

$$P\,m\,n_0 = 2p_{N_2} \tag{A10}$$

and, from Equation A2:

$$P\,n_0 = p_{CO} + 2p_{CO_2} + 2p_{O_2} + p_{H_2O} \tag{A11}$$

Substituting Equations A8, A9 and A10 in Equation A2 gives:

$$n_0 = \frac{(1 - p_{O_2}/P)}{k + \ell/2 + m/2} = \frac{(1 - p_{O_2}/P)}{z} \tag{A12}$$

where $z = k + \ell/2 + m/2$.

Setting $c = p_{O_2}$ in order to write the following equations more conveniently, the iteration used is:

$$c_{n+1} = c_n^2/P - \left[\frac{k}{(K_1/c^{\frac{1}{2}}) + 1} + \frac{\ell/2}{(K_2/c^{\frac{1}{2}}) + 1} + k - 1\right]\left[\frac{(1 - c_n/P)^2}{2z}\right] \tag{A13}$$

After iteration:

$$n_0 = (1 - c/P)/z \tag{A14}$$

$$p_{CO} = Pkn_0K_1/(K_1 + c^{\frac{1}{2}}) \tag{A15}$$

$$p_{CO_2} = p_{CO}\, c^{\frac{1}{2}}/K_1 \quad \text{or} \quad Pkn_0 - p_{CO} \tag{A16}$$

$$p_{H_2} = P\ell n_0\, K_2/(2K_2 + c^{\frac{1}{2}}) \tag{A17}$$

$$p_{H_2O} = p_{H_2}\, c^{\frac{1}{2}}/K_2 \tag{A18}$$

and

$$p_{N_2} = Pmn_0/2 \tag{A19}$$

The calculation comprises the following steps:

(1) Estimate the partial pressure of oxygen.

(2) Perform the iteration given in Equation A13.

(3) Calculate the remaining partial pressures.

For example, the calculation of the products of combustion of methane in air at atmospheric pressure yields the following data for products at 2000 K.

$K_1 = 1.306 \times 10^{-3}$ $\quad$ $K_2 = 2.884 \times 10^{-4}$

Initial guess	0.05
1st iteration	2.86×10^{-3}
2nd iteration	1.64×10^{-3}
3rd iteration	2.15×10^{-3}
4th iteration	1.89×10^{-3}
5th iteration	2.01×10^{-3}
6th iteration	1.95×10^{-3}
7th iteration	1.98×10^{-3}
8th iteration	1.96×10^{-3}
$p_{CO} \times 100$	0.271
$p_{CO_2} \times 100$	9.22
$p_{O_2} \times 100$	0.197
$p_{H_2} \times 100$	0.123
$p_{H_2O} \times 100$	18.85
$p_{N_2} \times 100$	71.34

The data show that, although the initial estimate was 25 times larger than the true value, convergence was obtained. Even an absurdly high initial guess of 0.5 gives convergence to 1.97×10^{-3} in 10 iterations so the method is certainly insensitive to the initial guess although convergence is not as rapid as in other techniques. This is, of course, offset by the simplicity of the iteration. In addition, succeeding values in the iteration oscillate about the true value and, if iteration is being done by hand, it is easy to interpolate and ensure very rapid convergence. The method is more sensitive to the initial estimate for the products of substoichiometric combustion but it is still very convenient even for this case.

(3) MORE EXTENSIVE DISSOCIATION

The concentrations of the species present in combustion products are calculated from a number of simultaneous equations which are of two kinds: equilibrium equations such as Equations A3 and A4 and conservation equations (Equations A2, A8, A9 and A10). For *n* species, *n* simultaneous equations are required, some of which (the equilibrium equations) are non-linear. Weinberg (1957) has devised a technique whereby a smaller number of linear simultaneous equations are derived, a procedure which simplifies the computation considerably when a large number of species in the products are considered. The derivation and method of calculation for nine species are set out below.

(a) Derivation

Basic units: carbon, hydrogen, oxygen and nitrogen atoms

Species present in the combustion products:	H_2	O_2	CO_2	CO	H_2O	OH	H	O	N_2
	(1)	(2)	(3)	(4)	(5)	(6)	(7)	(8)	(9)

Dissociation is described by:

$$CO_2 \rightleftharpoons CO + \tfrac{1}{2}\,O_2 \qquad \text{(A20)}$$

$$H_2O \rightleftharpoons H_2 + \tfrac{1}{2}\,O_2 \qquad \text{(A21)}$$

$$H_2O \rightleftharpoons \tfrac{1}{2}H_2 + OH \text{ or } \tfrac{1}{2}H_2 + \tfrac{1}{2}O_2 \rightleftharpoons OH \qquad \text{(A22)}$$

$$\tfrac{1}{2}H_2 \rightleftharpoons H \qquad \text{(A23)}$$

$$\tfrac{1}{2}O_2 \rightleftharpoons O \qquad \text{(A24)}$$

All species can be expressed in terms of three selected species, for example, p_{CO_2}, p_{H_2O} and p_{O_2}, by means of the equilibrium equations:

$$p_{CO} = K_1\, p_{CO_2}/p_{O_2}^{\frac{1}{2}} \qquad (A25)$$

$$p_{H_2} = K_2\, p_{H_2O}/p_{O_2}^{\frac{1}{2}} \qquad (A26)$$

$$p_{OH} = K_3\, p_{H_2O}/p_{H_2}^{\frac{1}{2}} = K_3(p_{H_2}\cdot p_{O_2})^{\frac{1}{2}}/K_2 \qquad (A27)$$

$$p_{H} = K_4\, p_{H_2}^{\frac{1}{2}} \qquad (A28)$$

$$p_{O} = K_5\, p_{O_2}^{\frac{1}{2}} \qquad (A29)$$

The conservation equations are expressed in terms of the number of atoms and the ratios of hydrogen to carbon atoms (R_{HC}) and oxygen to carbon (R_{OC}):

$$2p_{H_2O} + 2p_{H_2} + p_{OH} + p_H - R_{HC}(p_{CO_2} + p_{CO}) = 0 \qquad (A30)$$

$$2p_{CO_2} + p_{CO} + p_{H_2O} + 2p_{O_2} + p_{OH} + p_O - R_{OC}(p_{CO_2} + p_{CO}) = 0 \qquad (A31)$$

$$p_{CO_2} + p_{CO} + p_{H_2O} + p_{O_2} + p_{H_2} + p_{OH} + p_H + p_O + p_{N_2} - P = 0 \qquad (A32)$$

Since the number of molecules of nitrogen in air is 3.76 times that of oxygen:

$$p_{N_2} = 3.76(p_{CO_2} + p_{O_2}) + 1.88(p_{CO} + p_{H_2O} + p_{OH} + p_O) \qquad (A33)$$

and Equation A32 may be rewritten as:

$$4.76(p_{CO_2} + p_{O_2}) + 2.88(p_{CO} + p_{H_2O} + p_{OH} + p_O) + p_{H_2} + p_H - P = 0 \qquad (A34)$$

If the initial guesses of the three selected species are p'_{H_2O}, p'_{O_2} and p'_{CO_2} (y_1, y_2, y_3) the true values are:

$$p_{H_2O} = p'_{H_2O} + \delta_{H_2O} = y_1 + x_1 \quad \text{(A35)}$$

$$p_{O_2} = p'_{O_2} + \delta_{O_2} = y_2 + x_2 \quad \text{(A36)}$$

$$p_{CO_2} = p'_{CO_2} + \delta_{CO_2} = y_3 + x_3 \quad \text{(A37)}$$

where δ and x refer to small corrections.

Rewriting the conservation equations, A11, A12 and A13 by substituting $y + x$ for p (Equations A14, A15 and A16):

$$2x_1 + 2x_5 + x_{OH} + x_H - R_{HC}(x_3 + x_4) = z_1 \quad \text{(A38)}$$

$$2x_3 + x_4 + x_5 + 2x_2 + x_6 + x_8 - R_{OC}(x_3 + x_4) = z_2 \quad \text{(A39)}$$

$$4.76(x_3 + x_2) + 2.88(x_4 + x_5 + x_6 + x_8) + x_1 + x_7 = z_3 \quad \text{(A40)}$$

where

$$z_1 = -\left[2y_1 + 2y_5 + y_6 + y_7 - R_{HC}(y_3 + y_4)\right]$$

$$z_2 = -\left[2y_3 + y_4 + y_5 + 2y_2 + y_6 + y_8 - R_{OC}(y_3 + y_4)\right]$$

$$z_3 = -\left[4.76(y_3 + y_2) + 2.88(y_4 + y + y_6 + y_8) + y_1 + y_7 - P\right]$$

$x_4(\delta_{CO})$, $x_5(\delta_{H_2O})$, x_6, x_7 and x_8 can be expressed in terms of $x_1(\delta_{H_2})$, $x_2(\delta_{O_2})$ and $x_3(\delta_{CO_2})$ in functions like:

$$\delta_{CO} = \left(\frac{\partial p_{CO}}{\partial p_{CO_2}}\right)\delta_{CO_2} - \frac{1}{2}\left(\frac{\partial p_{CO}}{\partial p_{O_2}}\right)\delta_{O_2}$$

and

$$\delta_{H_2O} = \left(\frac{\partial p_{H_2O}}{\partial p_{H_2}}\right)\delta_{H_2} - \frac{1}{2}\left(\frac{\partial p_{H_2O}}{\partial p_{O_2}}\right)\delta_{O_2}$$

Weinberg has shown that $\partial p_{CO}/\partial p_{CO_2}$ can be set equal to p'_{CO}/p'_{CO_2} or y_4/y_3 etc. and so:

$$x_4 = (y_4/y_3)x_3 - \tfrac{1}{2}(y_4/y_2)x_2 \qquad \text{(A41)}$$

$$x_5 = (y_5/y_1)x_1 - \tfrac{1}{2}(y_5/y_2)x_2 \qquad \text{(A42)}$$

Equation A41 and A42 can be written directly by comparison with Equations A20 and A21. Similarly, inspection of Equations A22, A23 and A24 gives:

$$x_6 = \tfrac{1}{2}(y_6/y_1)x_1 + \tfrac{1}{2}(y_6/y_2)x_2 \qquad \text{(A43)}$$

$$x_7 = \tfrac{1}{2}(y_7/y_1)x_1 \qquad \text{(A44)}$$

$$x_8 = \tfrac{1}{2}(y_8/y_2)x_2 \qquad \text{(A45)}$$

Equations A38, A39 and A40 can be rewritten as:

$$a_{11}\, x_1 + a_{12}\, x_2 + a_{13}\, x_3 = z_1 \qquad \text{(A46)}$$

$$a_{12}\, x_1 + a_{22}\, x_2 + a_{23}\, x_3 = z_2 \qquad \text{(A47)}$$

$$a_{31}\, x_1 + a_{32}\, x_2 + a_{33}\, x_3 = z_3 \qquad \text{(A48)}$$

where

$$a_{11} = 2 + (2y_5 + y_6/2 + y_7/2)/y_1$$

$$a_{12} = (y_5 + y_6/2 + R_{HC}\, y_4/2)/y_2$$

$$a_{13} = -R_{HC}\,(1 + y_4/y_3)$$

$$a_{21} = (y_5 + y_6/2)/y_1$$

$$a_{22} = 2 + \left[(R_{OC} - 1)y_4 + y_5 + y_6 + y_8\right]/2y_2$$

$$a_{23} = 1 + (1 - R_{OC})(1 + y_4/y_3)$$

$$a_{31} = 1 + \left[y_7/2 + 2.88(y_5 + y_6/2)\right]/y_1$$

$$a_{32} = 4.76 + 2.88(y_5 - y_4 + y_6 + y_8)/2y_2$$

$$a_{33} = 4.76 + 2.88\, y_4/y_3$$

(b) Calculation

(1) Estimate the partial pressures of the selected species.

(2) Calculate the corresponding pressures of the other species (Equations A25 to A29).

(3) Evaluate the coefficients in the linear simultaneous equations (A25, A26 and A27).

(4) Solve for errors (x_1, x_2, x_3).

(5) Go to step 2 and repeat until errors are negligible.

The Weinberg technique can also be extended to the computation of a much larger number of species. For example, Gray and Smith (1967) used it for the estimation of 21 species which included, by use of the Saha equation, K^+ ions and electrons, with the object of estimating the conductivity of the combustion products.

When the partial pressures of the species in the combustion products are determined directly, as in the Weinberg method, the total number of moles derived from 1 mole of fuel of formula $C_u H_v$ is

$$\frac{uP}{p_{CO_2} + p_{CO}} .$$

The Weinberg technique, however, requires reasonably good guesses of the selected species. A less demanding approach in this respect is based on the use of n simultaneous equations for n species. Dixon-Lewis and Greenberg (1975) have described a calculation of this kind based on the Newton Raphson iterative technique.

An alternative technique is the minimisation of the free energy of the mixture of products by altering its composition without altering the atom balance (Van Zeggeren and Storey, 1970). Its advantage lies in its lack of sensitivity to the initial guesses of the concentration of the components; its drawback is that it requires more computational effort than the equilibrium technique. The methods involving equilibrium constants and the minimisation of free energy are, of course, entirely equivalent.

Appendix 2 Flame Temperatures

The combustion of fuel and oxidant at a temperature T_1 to produce combustion products at an adiabatic flame temperature T_2 may be represented by Figure A.1.

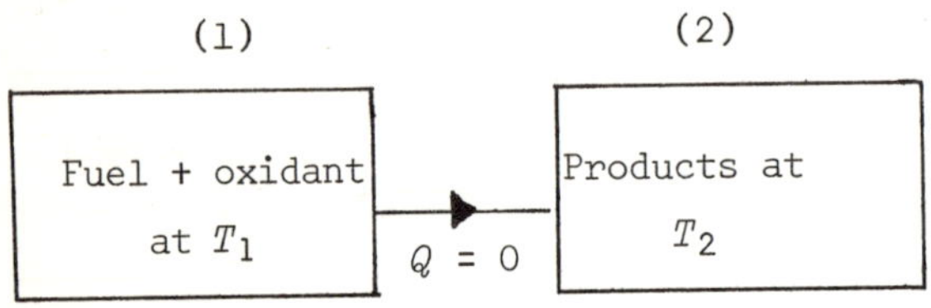

Figure A.1

Q is the heat transferred to the surroundings. The temperature T_2 is determined, in principle, by the changes of enthalpy which take place on combustion and those that correspond to the maintenance of the products of combustion at T_2. However, these changes cannot be obtained from the single step shown above and the calculation is made using steps for which the changes of enthalpy are known. Implicit in this approach is the fact that changes in enthalpy do not depend on the path of a process but only on the initial and final states. Consequently, the steps in combustion can be represented by (b), (c) and (d) in Figure A.2.

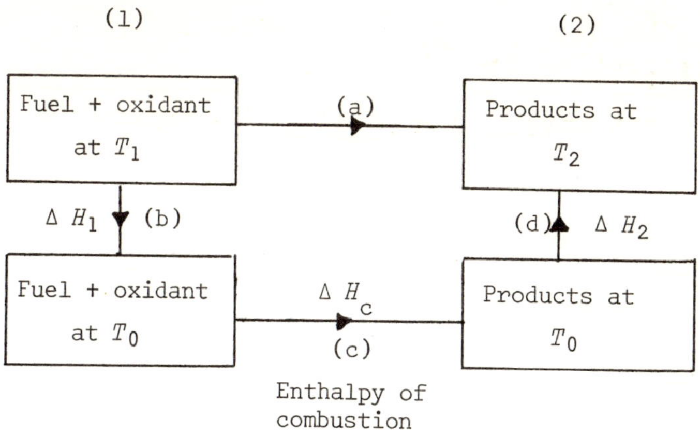

Figure A.2

T_0 is a standard temperature, such as 25°C, chosen because the enthalpy of combustion for pure compounds is usually known at this temperature. Alternatively, the calorific value may be used here if a pure compound is not being burned.

The change in enthalpy for route (a) is the same as that for (b), (c) and (d).

$$Q = 0 = H_2 - H_1 \quad \text{(A49)}$$

$$= \Delta H_1 + \Delta H_c + \Delta H_2 \quad \text{(A50)}$$

The value of ΔH_c or the calorific value (CV) is negative, and

$$CV = \Delta H_1 + \Delta H_2 \quad \text{(A51)}$$

$$\text{or} \quad CV - (\Delta H_1 + \Delta H_2) = 0 \quad \text{(A52)}$$

ΔH_1 is often expressed in terms of specific heat multiplied by the temperature difference $(T_0 - T_1)$, whereas ΔH_2 is either expressed in terms of the specific heats of the products or the enthalpies at T_2. If expressed as specific heats, an equation involving T_2 is produced; if, as enthalpies, a number of calculations for selected temperatures (T) must be made and T_2 determined by interpolation (see Figure A.3)

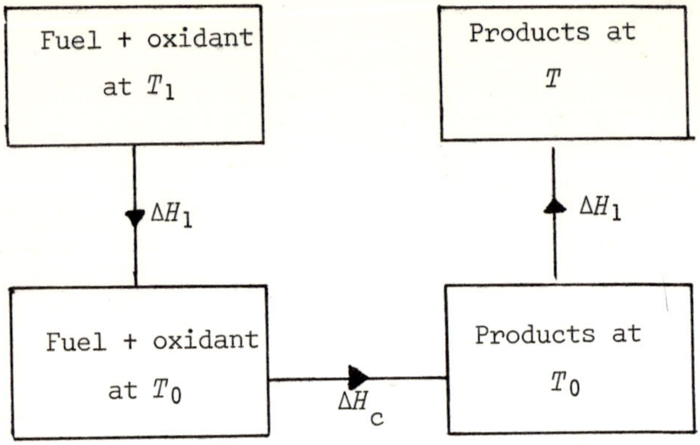

Figure A.3

If T equals T_2, CV - $(\Delta H_1 + \Delta H) = 0$. If T is not the adiabatic flame temperature:

$$\mathrm{CV} - \Delta H + \Delta H_T = \Delta_T$$

A plot of Δ_T against T will give an intercept on the T axis which corresponds to the flame temperature (Figure A.4).

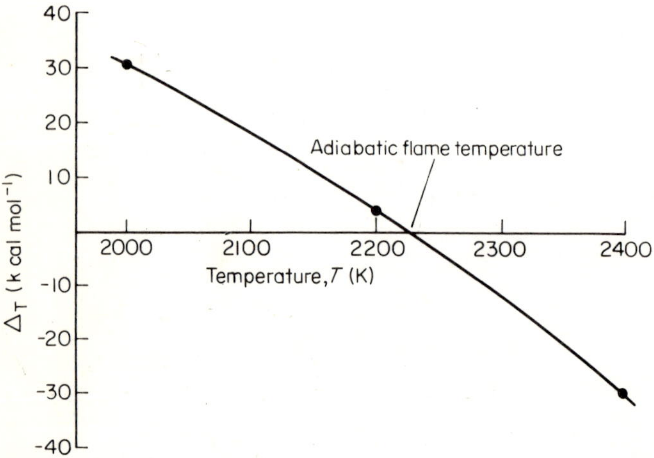

Figure A.4 Δ_T *versus* T

DISSOCIATION

Dissociation is taken into account by incorporating enthalpy changes of dissociation into the scheme given in Figure A.3, giving the scheme represented in Figure A.5.

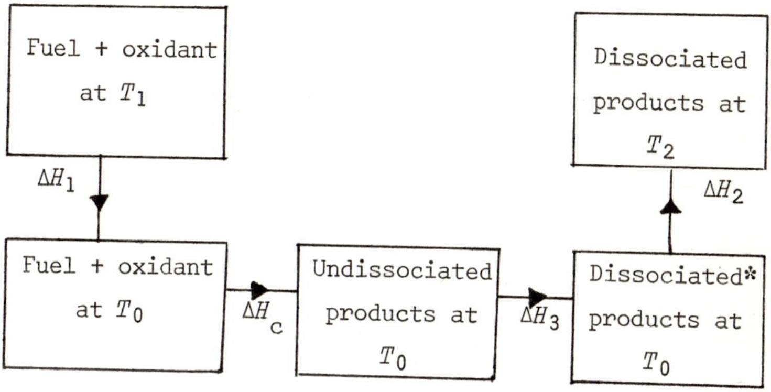

Figure A.5. (*in this scheme, the products are dissociated to an extent dictated by the temperature T_2; ΔH_2 then refers merely to the heating of these products from T_0 to T_2.)

ΔH_3 is the enthalpy of dissociation at the standard temperature, T_0.

Summarising, the methods outlined above for determining the adiabatic flame temperature incorporate two steps: (1) determination of the composition of the products of combustion at temperature T_2 and (2) an enthalpy balance over the steps required to effect the transition from fuel + oxidant at T_1 to products at T_2. The following example shows the determination of the adiabatic flame temperature of methane in air (at 25°C) in stoichiometric proportions by making use of the composition of the products of combustion at temperatures 2000, 2200 and 2400 K. The pressure is atmospheric.

Table A2.1 Composition of the products of combustion of methane in air (stoichiometric)

Component	2000 K	2200 K	2400 K
N_2	0.713	0.709	0.702
H_2O	0.188	0.184	0.177
CO_2	0.0918	0.0862	0.0760
CO	0.00307	0.00820	0.0176
O_2	0.00168	0.00428	0.00886
H_2	0.00135	0.00330	0.00700
OH	0.00072	0.00223	0.00558
NO	0.00069	0.00179	0.00387
O	0.00003	0.00019	0.00088
H	0.00006	0.00032	0.00133

The enthalpy changes on dissociation (ΔH_3) can be determined by setting up a series of equations which describe the production of each component:

$$CO_2 \rightarrow CO + \tfrac{1}{2}O_2 + 67.737 \text{ kcal mol}^{-1}$$

(derived from JANAF data)

$$H_2O \rightarrow H_2 + \tfrac{1}{2}O_2 + 57.798 \text{ kcal mol}^{-1}$$

$$\tfrac{1}{2}H_2 + \tfrac{1}{2}O_2 \rightarrow OH + 9.432 \text{ kcal mol}^{-1}$$

$$\tfrac{1}{2}H_2 \rightarrow H + 52.102 \text{ kcal mol}^{-1}$$

$$\tfrac{1}{2}O_2 \rightarrow O + 59.559 \text{ kcal mol}^{-1}$$

$$\tfrac{1}{2}N_2 + \tfrac{1}{2}O_2 \rightarrow NO - 21.58$$

$$\Delta H_{3,T} = N\Big[(67.637p_{CO} + 57.798(p_{H_2} + p_H/2 + p_{OH}/2) + $$
$$+ 9.43p_{OH} + 52.102p_H + 59.559p_O - 21.58p_{NO})\Big]/P$$

where N is the total number of moles of the products and p refers to the partial pressure of each component corresponding to temperature T. P is the total pressure which is equal to 1 in this case.

$$\Delta H = N(p_{N_2}\,\Delta H_{N_2} + p_{H_2O}\,\Delta H_{H_2O} + p_{CO_2}\,\Delta H_{CO_2} + p_{O_2}\,\Delta H_{O_2} + p_{OH}\,\Delta H_{OH} + p_{NO}\,\Delta H_{NO} + p_{O}\,\Delta H_{O} + p_{H}\,\Delta H_{H})$$

where ΔH refers to the enthalpy change on heating from T_0 to T.

$$\Delta H_c = 191.76 \text{ kcal mol}^{-1}$$

$\underline{T = 2000 \text{ K}}$

$$\Delta_{2000} = \Delta H_c - \Delta H_1 - \Delta H_{2000} - \Delta H_{3,2000}$$

$$= 191.76 - 0 - 157.51 - 3.53$$

$$= 30.72 \text{ kcal mol}^{-1}$$

$\underline{T = 2200 \text{ K}}$

$$\Delta_{2200} = \Delta H_c - \Delta H_1 - \Delta H_{2200} - \Delta H_{3,2200}$$

$$= 191.76 - 0 - 178.33 - 9.6$$

$$= 3.8 \text{ kcal mol}^{-1}$$

$\underline{T = 2400 \text{ K}}$

$$\Delta_{2400} = \Delta H_c - \Delta H_1 - \Delta H_{2400} - \Delta H_{3,2400}$$

$$= 191.76 - 0 - 199.46 - 21.93$$

$$= -29.63 \text{ kcal mol}^{-1}$$

Plotting Δ_T against T in Figure A.4 gives the flame temperature as 2225 K. It is possible to combine the steps as Dixon-Lewis and Greenberg (1975) have done by incorporating the flame temperature as a variable in the equilibrium calculation.

TOTAL ENTHALPIES

An alternative, but entirely equivalent, technique for determining the flame temperatures of pure compounds is based on the statement that the total enthalpy of the products of combustion minus the total enthalpy of the reactants equals the heat transferred during combustion (Q). For the adiabatic flame temperature, $Q = 0$, and

$$\text{total enthalpy of products} - \text{total enthalpy of reactants} = 0 \qquad \text{(A53)}$$

The total enthalpy for each species is the enthalpy of formation at 25°C plus the sensible enthalpy relative to 25°C.

The calculation, by this technique, of the adiabatic flame temperature of methane in air at 25°C in stoichiometric proportions requires the total enthalpies of the products (Table A2.2) as well as the composition of the products at temperatures in the vicinity of the flame temperature (Table A2.1). At a temperature other than the adiabatic flame temperature:

$$\text{total enthalpy of products} - \text{total enthalpy of reactants} = \Delta_{T} \qquad \text{(A54)}$$

The adiabatic flame temperature ($\Delta H_T = 0$) is determined by interpolation from the values of ΔH_T for temperatures in the vicinity of the flame temperature.

2000 K

$$\Delta_{2000} = -30.72 \text{ kcal mol}^{-1}$$

2200 K

$$\Delta_{2200} = -3.6 \text{ kcal mol}^{-1}$$

2400 K

$$\Delta_{2400} = 29.48 \text{ kcal mol}^{-1}$$

The flame temperature is 2225 K which is the value obtained using the enthalpies of dissociation.

Table A2.2 Total enthalpies of the reactants and products of the stoichiometric combustion of methane at 25°C and atmospheric pressure (from data from the JANAF tables). The enthalpy of formation of methane is -17.895 kcal mol^{-1}

Component	2000 K	2200 K	2400 K
N_2	13.42	15.15	16.89
H_2O	-40.43	-37.95	-35.43
CO_2	-72.19	-69.29	-66.38
CO	-12.86	-11.12	- 9.37
O_2	14.15	15.97	17.80
H_2	12.65	14.31	15.99
OH	22.28	23.95	25.64
NO	35.41	37.17	38.93
O	68.10	69.09	70.09
H	60.56	61.55	62.54
Moles of products per mole of CH_4	10.54	10.59	10.68

Units and Nomenclature

UNITS

m	metre (1Å = 10^{-10} m)	or standard multiples or submultiples of these units
kg	kilogramme	
s	second	
K	degree kelvin	
J	joule	
W	watt or J s^{-1}	
rad	radian	
ster	steradian	
mol	gram mole	

NOMENCLATURE

A	area (m^2)
A_m	projected area per unit mass of particle material (m^2 kg^{-1})
A_n	projected area per particle (m^2)
A_v	projected area per unit volume of particle material (m^2 m^{-3})
C_p	molar heat capacity (J mol^{-1} K^{-1})
C_{CO_2}	correction factor for the emissivity of carbon dioxide

C_{H_2O} correction factor for the emissivity of water vapour

D diameter, characterising dimension (m)

D_{ab} diffusion coefficient for component a in component or components b

$D_{i\text{-mix}}$ diffusion coefficient for component i in gaseous mixture

E radiative flux emitted by a body (W m^{-2})

E_b radiative flux emitted by a black body (black body emissive power)(W m^{-2})

E_λ monochromatic radiative flux emitted by a body (W m^{-3})

$E_{b\lambda}$ monochromatic black body emissive power (W m^{-3})

F_{ab} relative absorption cross-sectional area

F_{at} relative attenuation cross-sectional area

F_{sc} relative scattering cross-sectional area

F_{12} view factor from surface 1 to surface 2

G mass flux (kg m^{-2}s^{-1})

G_r Grashof number ($\beta\ g\ \rho^2\ L^3\ \Delta T/\mu^2$)

H radiative flux incident upon a surface (W m^{-2}); molal enthalpy (J mol^{-1})

I intensity of radiation (W m^{-2} ster^{-1})

I_n intensity of radiation normal to a surface (W m^{-2} ster^{-1})

K attenuation or extinction coefficient (m^{-1})

L length (m)

L_m mean beam length (m)

L_{12} mean beam length between surface (or medium) 1 and surface (or medium) 2 (m)

Le Lewis number ($\rho\ D_{i\text{-mix}}\ c_p/\lambda$)

M molecular weight (g mol^{-1})

Nu	Nusselt number ($h\ D/k$)
P	total pressure (bar = 10^5 N m^{-2})
Pr	Prandtl number ($c_p\ \mu/k$)
Q	rate of heat transfer (W)
R	random number; gas constant
Re	Reynolds number ($\rho\ U\ D/\mu$)
Sc	Schmidt number ($\mu\ \rho\ D_{i\text{-mix}}$)
S_m	surface area of particles per unit mass of particle material ($m^2\ kg^{-1}$)
S_n	surface area of a particle (m^2)
St	Stanton number ($h/c_p\ \rho\ U$)
S_v	surface area of particles per unit volume of particle material ($m^2\ m^{-3}$)
T	absolute temperature (K)
T_b	effective black body (or brightness) temperature (K)
U	velocity ($m\ s^{-1}$)
V	volume (m^3)
W	radiative flux leaving a surface (radiosity)($W\ m^{-2}$)
$W(T)$	Wien approximation for $E_{b\lambda}$($W\ m^{-3}$)
$\overline{12}$	exchange area between two surfaces (1 and 2)(m^2)
a	weighting coefficient for grey gas fit
c_m	mass concentration of particles in suspension ($kg\ m^{-3}$)
c_n	number of particles per unit volume of suspension
c_p	specific heat at constant pressure ($J\ kg^{-1}\ K^{-1}$)
c_v	volume concentration of particles in suspension ($m^3\ m^{-3}$)

c_1, c_2	constants in Planck's expression
d	particle diameter (m)
$\overline{g_1 g_2}$	exchange area between two gases (1 and 2) (m^2)
$\overline{g_1 s_2}$	exchange area between a gas (1) and a surface (2) (m^2)
h	enthalpy per unit mass ($J\ kg^{-1}$)
h_c	convective heat transfer coefficient ($W\ m^{-2}\ K^{-1}$)
h_D	convective mass transfer coefficient ($m\ s^{-1}$)
k	thermal conductivity ($W\ m^{-1}\ K^{-1}$)
k	attenuation coefficient used in conjunction with partial pressure ($m^{-1}\ bar^{-1}$)
k'	absorption index
ℓ	length (m)
m	complex optical property
m	mass (kg)
n	refractive index
p	partial pressure (bar = $10^5\ N\ m^{-2}$)
q	rate of heat transfer per unit area ($W\ m^{-2}$)
r	radius (m)
$\overline{s_1 s_2}$	exchange area between two surfaces (1 and 2) (m^2)
t	time (s)
v	velocity ($m\ s^{-1}$)
w	solid angle (ster)
x_i	mole fraction of component i
α	absorptivity
α_λ	monochromatic absorptivity
β	coefficient of cubical expansion (K^{-1})
ε	emissivity
ε_g	emissivity of a gas

ε_λ	monochromatic emissivity
η	efficiency
θ	circumferential angle (rad)
λ	wavelength (m)
λ_{max}	wavelength at which maximum flux is emitted (m)
μ	absolute viscosity (N s m^{-2})
ν	frequency (s^{-1})
ρ	density (kg m^{-3})
ρ	reflectivity
ρ_λ	monochromatic reflectivity
σ	Stefan Boltzmann constant (W m^{-2} K^{-4})
τ	transmissivity
τ_λ	monochromatic transmissivity
ϕ	angle between direction of emission and normal to the surface (cone angle) (rad)

References

Arthur, J.R. (1950) *Nature*, 165, 557.

Brokaw, R.S. (1958) *J. Chem. Phys.*, 29, 391.

Chilton, T.H. and Colburn, A.P. (1934) *Ind. Eng. Chem.*, 26, 1183.

Davies, R.M., Lucas, D.M. and Toth, H.E. (1971) *Gas Council Res. Comm.* Gc 184.

Davies, R.M. and Oeppen, B. (1972) *J. Inst. Fuel*, 45, 383.

Dixon-Lewis, G. and Greenberg, J.B. (1975) *J. Inst. Fuel*, 48, 132.

Dunkle, R.V. (1964) *J. Heat Transfer* (ASME), 75.

Field, M.A., Gill, D.W., Morgan, B.B. and Hawksley, P.G.W. (1967) 'Combustion of pulverized coal', *Br. Coal Util. Res. Assoc., Man. Bull*, available from Institute of Fuel, London.

Francis, W.E., Moppett, B.E. and Read, G.P. (1967) *J. Inst. Gas Eng.*, 7, 335.

Gaydon, A.G. and Wolfhardt, H.G. (1970) *Flames*, Chapman and Hall, London.

Golothan, D.W. (1967) *Soc. Auto. Eng. Trans.*, 76, 616.

Gray, W.A. and Muller, R. (1974) *Engineering Calculations in Radiative Heat Transfer*, Pergamon, Oxford.

Gray, W.A. and Smith, J.L. (1967) *J. Inst. Fuel*, 40, 187.

Hadvig, S. (1970) *J. Inst. Fuel*, 43, 129.

Harker, J.H. (1967) *J. Inst. Fuel*, 40, 206.

Hirschfelder, J.O. (1957) *J. Chem. Phys.*, 26, 274.

Hirschfelder, J.O., Curtiss, C.Y. and Bird, R.B. (1954) *Molecular Theory of Gases and Liquids*, Wiley, New York.

Hottel, H.C. and Sarofim, A.F. (1967) *Radiative Transfer*, McGraw Hill, New York.

Joint Army, Navy and Air Force (J.A.N.A.F.) (1965) *Thermochemical Tables*, Clearing House for Federal, Scientific and Technical Information, PB 168, 370.

Kilham, J.K. and Cookson, R.A. (1963) Ninth Symposium (International) on Combustion p. 257, Academic Press, New York.

Kilham, J.K. and Dunham, P.G. (1967) Eleventh Symposium (International) on Combustion p. 899, The Combustion Institute, Pittsburgh.

Kilham, J.K. and Purvis, M.J.J. (1971) *Combustion and Flame*, 16, 47.

Kolenda, Z. and Gray, W.A. (1974) *Bull. Pol. Acad. Sci.*, 22, 25.

Kreith, F. (1962) *Radiation Heat Transfer for Spacecraft and Solar Plant*, International Textbook, Scranton.

Kreith, F. (1973) *Principles of Heat Transfer*, Inter. Textbook Publishers, Scranton.

Lees, L. (1956) *Jet Propulsion*, 26, 259.

Leuenberger, H. and Person, R.A. (1956) ASME Paper No. 56-A144.

Littlejohn, R.F. (1966) *J. Inst. Fuel*, 39, 59.

Mason, E.A. and Saxena, S.C. (1958) *Phys. of Fluids*, 1, 361.

McAdams, W.H. (1954) *Heat Transmission*, McGraw Hill, New York.

McCartney, J.J. and Ergun, S. (1958) *Fuel*, 37, 272.

Mitchell, S.C., (1976) *J. Inst. Fuel*, 49.

Parker, W.G. and Wolfhard, H.G. (1950) *J. Chem Soc.*, 2038.

Rose, H.E. (1952) *J. Appl. Chem.*, 2, 80.

Rosner, D. (1961) *Aeron Res. Lab. Rep.* ARL99, Part 1, AD269816.

Sarofim, A.F. (1962) Sc.D. thesis in Chem. Eng., M.I.T., Cambridge, Mass.

Sparrow, E.M. and Cess, R.D. (1966) *Radiation Heat Transfer*, Brooks/Cole, California.

Street, J.C. and Thomas, A. (1955) *Fuel*, 34, 5.

Strehlow, R.A. (1968) *Fundamentals of Combustion*, Int. Textbook Company, Scranton.

Tesner, P.A. (1959) 7th International Symposium on Combustion, 546.

Thomas, A. (1962) *Combustion and Flame*, 6, 46.

Thring, M.W. (1962) *The Science of Flames and Furnaces*, Chapman and Hall, London.

Trinks, W. and Mawhinney, M.H. (1961) *Industrial Furnaces*, Wiley, New York.

Van Zeggeren, F. and Storey, S.H. (1970) *Computation of Chemical Equilibria*, Cambridge University Press, London.

Wiebelt, J.A. (1966) *Engineering Radiation Heat Transfer*, Holt, Rinehart and Winston, New York.

Wilke, C.R. (1950) *J. Chem Phys.*, 18, 517.

Winding, C.C. and Cheney, A.J. (1948) *Ind. Eng. Chem.*, 40, 1087.

Weinburg, F.J. (1957) *Proc. Roy. Soc. A.*, 241. 132.

Weinburg, F.J. (1968) *Proc. Roy. Soc. A.*, 307, 195.

Index